TEXTBOOK OF RESEARCH ETHICS
Theory and Practice

Sana Loue
Case Western Reserve University
Cleveland, Ohio

KLUWER ACADEMIC / PLENUM PUBLISHERS
New York, Boston, Dordrecht, London, Moscow

Appendix 1 appeared in part as Chapter 3 in Loue, *Forensic Epidemiology*,
©1999 Board of Trustees, Southern Illinois University Press, reprinted with permission.

All figures ©1999 Hal Morgenstern, Ph.D., reprinted with permission.

ISBN 0-306-46448-9

©2000 Kluwer Academic / Plenum Publishers, New York
233 Spring Street, New York, New York 10013

http://www.wkap.nl/

10 9 8 7 6 5 4 3 2 1

A C.I.P. record for this book is available from the Library of Congress

All rights reserved

No part of this book may be reproduced, stored in a retrieval system, or transmitted in any form
or by any means, electronic, mechanical, photocopying, microfilming, recording, or otherwise,
without written permission from the Publisher

Printed in the United States of America

A long habit of not thinking a thing wrong gives it a superficial appearance of being right, and raises at first a formidable outcry in defense of custom.

Thomas Paine, *Common Sense*, 1776

PREFACE

The *International Ethical Guidelines for Biomedical Research Involving Human Subjects* (CIOMS and WHO, 1993: 11) defines "research" as referring to

> a class of activities designed to develop or contribute to generalizable knowledge. Generalizable knowledge consists of theories, principles or relationships, or the accumulation of information on which they are based, that can be corroborated by accepted scientific techniques of observation and inference.

The *International Guidelines for the Ethical Review of Epidemiological Studies* (CIOMS, 1991) recognizes that it may be difficult to distinguish between research and program evaluation. It offers the following guidance: "The defining attribute of research is that it is designed to produce new, generalizable knowledge, as distinct from knowledge pertaining only to a particular individual or programme" (CIOMS, 1991, Guideline 52, 23).

Health research includes both medical and behavioral studies that relate to health. Research can be conducted in conjunction with patient care (clinical research), or it can be conducted outside of the context of clinical care. Research may involve only observation, or it may require, instead or in combination, a physical, chemical, or psychological intervention. Research may generate new records or may rely on already-existing records.

Frank Press (Committee on the Conduct of Science, 1989: v) has observed that:

> [o]ne of the most appealing features of research is the great degree of personal freedom accorded scientists—freedom to pursue exciting opportunities, to exchange ideas freely with other scientists, to challenge conventional knowledge. Excellence in science requires such freedoms, and the institutions that support science in the United States have found ways to safeguard them. However, modern science, while strong in many ways, is also fragile in important respects....

One such respect relates to the lack of universal agreement regarding human rights and their implementation, specifically in the research context. Although international declarations and guidelines exist, they represent aspirational goals rather than a code designed to regulate conduct in specific situations (see Christakis and Panner, 1991). Even within the United States, research funded by states or private sources may not be subject to the standards imposed by federal regulations (Moreno, Caplan, Wolpe, and Members of the Project on Informed Consent, Human Research Ethics Group, 1998), potentially allowing the research participants to be deprived of important rights, whether through ignorance, negligence, or intent.

The consequences of such deficiencies and deprivations can be far-reaching. In addition to the physical, psychological, emotional, and/or economic harm that may befall individual participants, a belief in the integrity of science and scientists may be diminished or destroyed. Individuals approached for their participation in research may view such an undertaking as if they are but a means to the scientist's end, rather than an opportunity to "become a collaborator with investigators by effectively adopting the study's ends as [their] own" (Capron, 1991: 186).

This text is designed to assist in obviating those deficiencies and deprivations that may arise from ignorance. It provides a brief history of experimentation on humans and reviews various theories of ethics from which we have derived principles and rules that govern research involving human beings although, admittedly, some such principles and rules may not be universally accepted or their means of implementation may not be universal. Relevant international documents and national regulations, policies, and memoranda are referred to extensively to provide guidance in addressing issues that regularly arise in the course of research. The reader is challenged to examine research situations presenting ethical issues for resolution, to devise creative strategies for the integration of science, ethics, and, where relevant, law. And, to the extent that any one text can serve such a function, the text is a reminder that "concern for man himself must always form the chief interest of all technical endeavors...in order that the creations of our minds shall be a blessing and not a curse to mankind. Never forget this in the midst of your diagrams and equations" (Einstein, quoted in Committee on the Conduct of Science, 1989: 20).

REFERENCES

Capron, A.M. (1991). Protection of research subjects: Do special rules apply in epidemiology? *Law, Medicine & Health Care, 19*, 184-190.

Christakis, N.A. & Panner, M.J. (1991). Existing international ethical guidelines for human subjects research: Some open questions. *Law, Medicine & Health Care, 19*, 214-221.

Committee on the Conduct of Science, National Academy of Sciences. (1989). *On Being a Scientist*. Washington, D.C.: National Academy Press.

Council for International Organizations of Medical Sciences (CIOMS). *International Guidelines for Ethical Review of Epidemiological Studies*. Geneva: Author.

Council for International Organizations of Medical Sciences (CIOMS) & World Health Organization (WHO). (1993). *International Ethical Guidelines for Biomedical Research Involving Human Subjects*. Geneva: Author.

Moreno, J., Caplan, A.L., Wolpe, P.R., & Members of the Project on Informed Consent, Human Research Ethics group. Updating protection for human subjects involved in research. *Journal of the American Medical Association, 280*, 1951-1958.

ACKNOWLEDGMENTS

The author gratefully acknowledges the helpful review and critique of earlier versions of this manuscript by Drs. Stephen Post and Siran Karoukian. Dr. Hal Morgenstern is to be thanked for his willingness to permit the use of the figures which appear in the text. Many students contributed their time in hunting down obscure references and materials from other libraries and deserve praise and thanks for their time and diligence: Sandy Ferber, Jay Fiedler, Stephanie Stewart, and Fatoumata Traore. Mariclaire Cloutier, the editor of this volume, deserves praise for her patience and support.

CONTENTS

Chapter 1. Human Experimentation and Research: A Brief Historical Overview   1

   The Tuskegee Experiment   2
      The Historical and Social Context   2
      The Tuskegee Syphilis Study   6
   The Nazi Experiments   8
      Eugenics and Racial Hygiene   8
      The Recruitment of Physicians to the Nazi Cause   11
      The "Cleansing" of the Aryan "Body"   13
      Regulations Governing Human Experimentation   14
      The Medical Experiments   18
   The Cold War Experiments   19
      The Cold War: How It Began and What It Meant   19
      The Vanderbilt Nutrition Study   20
      The Fernald State School Experiments   21
      Total Body Irradiation at the University of Cincinnati   22
      The Aftermath   23
   The Prison Experiments   25
      The Holmesburg Prison Experiments   25
      Malaria, Leukemia, and Pellagra   28
   The Development of Diethylstilbesterol (DES)   30
   The Willowbrook Hepatitis Experiments   30
   The Tearoom Trade   31
   The Development of International Codes and Guidelines   32
      The Nuremberg Code and Declaration of Helsinki   32
      The International Covenant on Civil and Political Rights   33
      CIOMS and WHO International Guidelines   34
   The Development of United States Guidelines and Regulations   35
   Chapter Summary   37
   Exercise   37
   References   38

Chapter 2. Approaches to Ethical Analysis   45

   Casuistry   45
   Communitarianism   47
   Feminist Ethics   48
      Feminine or Feminist?   48
      Foundations of Feminine and Feminist Ethics   51
      Feminine Approaches to Ethics   51
      Feminist Approaches to Ethics   55

Principlism	58
Deontology	60
Utilitarianism	61
Contract-Based Ethics	64
Virtue Ethics	65
Pragmatism	65
Chapter Summary	66
Exercise	66
References	66

Chapter 3. Ethical Issues Before the Study Begins — 71

Study Design	71
Designing the Study	71
Formulating the Research Team	74
Formulating Inclusion and Exclusion Criteria	76
Working with Vulnerable Participants	77
Selecting a Comparison Treatment or Intervention	83
Balancing the Risks and Benefits	86
Ethical Review Committees	89
The Purpose and Function of the Review Committees	89
Ethical Review in the United States: The Institutional Review Board	91
One Institutional Review Board: An Example	95
Institutional Review and International Research	97
Exercise	99
Conflicts of Interest	99
Financial Conflicts	100
Other Conflicts	104
Strategies to Address Conflicts Before the Study Begins	105
Chapter Summary	108
Exercise	109
References	109

Chapter 4. Ethical Issues During and After the Study — 113

The Informed Consent Process	113
Recruitment	116
Explaining the Study	127
Designing the Informed Consent Process	142
Confidentiality and Privacy Concerns	147
Monitoring the Study	148
The Data Safety Monitoring Board	148
Community Consultation	149
Issues Relating to Confidentiality and Disclosure	149
Authorship	149
Obligation to Inform	159

Advocacy	162
Chapter Summary	163
References	164

Chapter 5. Legal Issues in Research — 171

Introduction	171
Misuse of Human Participants in Research	171
Scientific Misconduct	175
Defining Scientific Misconduct	175
Institutional Responses	179
Administrative Responses to Scientific Misconduct	182
Misconduct in Regulated Research	185
Legal Responses to Misuse of Human Participants, Scientific Misconduct, and Misconduct in Regulated Research	189
Civil Proceedings	189
Criminal Proceedings	193
Releasing Data in a Legal Context	196
Limits on Confidentiality	196
Mechanisms to Enhance Confidentiality	208
Chapter Summary	212
References	212
Appendix 1: Principles of Research Design	217
Appendix 2: Basic Legal Concepts	241
Index	249

LIST OF FIGURES

1.	Diagrammatic Representation of a Cohort Study	224
2.	Diagrammatic Representation of a Case-Control Study	225
3.	Diagrammatic Representation of a Cross-Sectional Study	227
4.	Diagrammatic Representation of an Ecological Study	228
5-19	Confounding	230

LIST OF TABLES

1.	Questions to Guide the Ethical Optimization of a Study	74
2.	Items to Be Addressed in the Preparation of an IRB Submission	92
3.	Considerations in the Development of an Informed Consent Process	117
4.	Considerations in Formulating Recruitment Strategies	124

TEXTBOOK OF
RESEARCH ETHICS

1
HUMAN EXPERIMENTATION AND RESEARCH: A BRIEF HISTORICAL OVERVIEW

> Young Soldier: I'm not going to listen. This is an injustice and injustice is something I will not endure.
>
> Mother Courage: Oh really? Gives you a problem, does it? Injustice? How long can you put up with it for? Is an hour difficult? Or does it bug you for two? Because in the sticks, I tell you, there's this strange sort of moment when people suddenly think, oh perhaps I can put up with injustice after all.
>
> Bertolt Brecht,
> Mother and Her Children

Health research involving human experimentation has been conducted throughout the world and, in the United States, for over a century. Ostensibly, research involving human experimentation has been subject to specified standards. For example, during the early years of the 20th century, humans were not to be used in experiments until after the safety of the new drug or procedure had been established in animals (Osler, 1907). Second, the "full consent" of a patient was a prerequisite to application of the new therapy. Patients entrusted to the care of the physician were not to be recruited for experimentation unless the new therapy would potentially result in a direct benefit to the individual patient. Third, the participation of healthy volunteers in experimentation was permissible, subject to the requirement of full knowledge of the circumstances and agreement to participate (Osler, 1907).

Despite these enunciated standards, research was not infrequently conducted under questionable circumstances. One experiment involving the injection of sterilized gelatin into two young boys and a "feeble-minded girl" resulted in "prostration and collapse" (Abt, 1903). Other researchers studying the ability of several new tests to detect tuberculosis injected tuberculin solution into more than 164 children under the age of 8, most of whom were residents of an orphanage. The experiment often resulted in discomfort, eye lesions, or eye inflammations (Belais, 1910; Hammill, Carpenter, and Cope, 1908). Other questionable experiments related to the etiology, diagnosis, and/or prevention of other diseases, including syphilis, yellow fever, typhoid, and herpes, were often conducted on children, prisoners, soldiers, or the mentally ill (Lederer, 1997).

This chapter focuses both on such research that has taken place in the United States and on research outside the United States that has had a significant impact on our conduct of research here. Only the most well-known and best-studied examples of human experimentation gone awry are presented. Although these occurrences are presented in chronological order, the time periods during which they occurred are often overlapping. The chapter also discusses the formulation and adoption of regulations in the United States and of international ethical guidelines in response to these experiments. The content of the guidelines and regulations is discussed more full throughout the text.

THE TUSKEGEE EXPERIMENT

One of the most infamous experiments in United States history has come to be known as the "Tuskegee experiment." An understanding of the historical and social context in which this experiment involving poor African Americans in a rural area of Alabama was conducted is critical to an understanding of how such an experiment could be initiated, how it could be allowed to continue, and how it has impacted and continues to impact human experimentation in the United States.

The Historical and Social Context

African-American Health During Slavery

Many physicians in the pre-Civil War South believed that significant medical differences existed between blacks and whites. Some physicians argued that blacks were immune from certain diseases that affected whites, such as malaria, but were especially susceptible to other conditions, such as frostbite (Savitt, 1985). One Northern physician observed of blacks that:

> God has adapted him, both in his physical and mental structure, to the tropics . . . . His head is protected from the rays of a vertical sun by a dense mat of wooly hair, wholly impervious to its fiercest heats, while his entire surface, studded with innumerable sebaceous glands, forming a complete excretory system, relieves him from all those climatic influences so fatal, under the same circumstances, to the sensitive and highly organized white man. Instead of seeking to shelter himself from the burning sun of the tropics, he courts it, enjoys it, delights in its fiercest heat. (Van Evrie, 1861: 251, 256)

African-Americans suffered from several causes of mortality, including pulmonary tuberculosis and neonatal tetanus, more so than whites. One particular form of tuberculosis, characterized by difficulty in breathing, abdominal pain, progressive debility and emaciation, and ultimately death, was so common among blacks that it became known as Negro Consumption or *Struma Africana*. Various explanations have been offered in attempts to explain the impact of this form of tuberculosis on blacks, including lack of an immune response to the disease due to lack of exposure and an increased susceptibility to serious first attacks of tuberculosis due to various factors, including malnutrition and pre-existing illness (Savitt, 1985).

Epidemics of cholera, yellow fever, and typhoid were of special concern among slaves. Slaves were often particularly vulnerable to cholera as a result of the increased consumption of water required by their strenuous work. The water, however, was often contaminated and the slaves frequently suffered from nutritional deficiencies that adversely affected their ability to recover from cholera (Lee and Lee, 1977).

The especial vulnerability to some diseases was used by some as an illustration of blacks' inferiority (Savitt, 1985). The harsh conditions to which black slaves were subjected were rarely mentioned as contributing to their susceptibility to specific diseases or to their poor health.

Male slaves were valued for their work, while the value of female slaves was determined not only by their capacity for work, but their capacity to reproduce and to increase the human property that formed the basis for the slave economy. Because female slaves were property, without any degree of autonomy, they were subject to their masters' sexual desires (Jacobs, 1988; Smith, 1988). Slaves were physically mutilated for real and imagined offenses (Hurmence, 1984; Jacobs, 1988). Lavinia Bell's treatment as a slave was all too common:

> After that time she was sent into the cotton field with the other field hands, where the treatment was cruelly severe. No clothes whatever were allowed them, their hair was cut off close to their head, and thus were exposed to the glare of a southern sun from early morn until late at night. Scarcely a day passed without their receiving fifty lashes, whether they worked or whether they did not. They were also compelled to go down on their knees, and harnessed to a plough, to plough up the land, with boys for riders, to whip them when they flagged in their work. At other times, they were compelled to walk on hackles, used for hackling flax. Her feet are now dotted over with scars, caused by their brutality . . . . Still later, for some disobedience on her part, they hoisted her into a tree, locked a chain round her neck, and hand-cuffed her wrists, the marks being yet visible. There she was left for two days and nights, without a morsel to eat, being taunted with such questions as to whether she was hungry and would like something to eat . . . (Blassingame, 1977: 342-343)

This failure to provide food for a slave was regarded as "the most aggravated development of meanness even among slaveholders" (Douglass, 1968: 34).

Disagreement exists with respect to the medical treatment that slaves received. Accounts from slaves seem to indicate that, even as judged by the standards of the time, medical care was often poor. Midwives or doctors were rarely in attendance at the birth of a slave's child. At least one author, however, has argued that medical care of slaves was often superior to that received by their owners, if only because the slave represented a financial investment which could be threatened by ill health (Kolchin, 1993). Slaves not infrequently resorted to remedies at home rather than report their illness to the person in charge and be required to submit to the medical care provided at the behest of the owner. Consequently, a dual system of health care developed (Savitt, 1985).

Poor living conditions exacerbated existing health problems. Although slaves were provided with housing, they were rarely provided with toilets (Blassingame, 1977). The housing itself was often characterized by poor ventilation, lack of light, and damp, earthen floors (Semmes, 1996). There was little opportunity to bathe or to wash clothes, resulting in the promotion of bedbugs and body lice. The soil and water were often infested with worms and larvae, to which the slaves were particularly vulnerable due to the lack of shoes and the poor sanitation (Blassingame, 1979; Savitt, 1978). Roundworm, threadworm, tapeworm, and hookworm infestations plagued many slaves. The practice of eating soil (geophagy), which was continued from West Africa, further promoted infestation with worms (Savitt, 1978).

Poor diets and food shortages further contributed to the development of poor health. Slaveowners frequently provided the slaves with pork, which was the preferred source of protein for the owners. However, the slaveowners retained the leanest cuts for themselves, and passed on the fatty portions, together with cornmeal, to the slaves. On some plantations, slaves rarely had dairy products, fruits, or vegetables (Stampp, 1956). Not surprisingly, the slaves' poor diet often resulted in deficiencies in vitamins, including vitamins A, B, C, and D. These deficiencies, in turn, led to diseases such as scurvy, beriberi, and pellagra (Savitt, 1978).

Slaves were the unwilling subjects of scientific experimentation. When compensation was offered, it was provided to the slaveowner. For example, the physician J. Marion Sims reached an agreement with one slaveowner to maintain several of his female slaves at Sims' expense in exchange for their use in experiments designed to repair vesico-vaginal fistulas (Sims, 1894).

African-American Health During Reconstruction

Poor housing and poor sanitation continued into the period of Reconstruction (Blassingame, 1973; Morais, 1967). African-Americans suffered from pellagra and other nutritional deficiencies (Johnson, 1966), for which they were held responsible. One physician opined

> His [the Negro] diet is fatty; he revels in fat; every pore of his sleek, contented face wreaks with unctuousness. To him the force-producing quality of the fats has the seductive fascination the opium leaves about the Oriental . . . (Tipton, 1886, quoted in Charatz-Litt, 1992: 717)

The high rates of death among African-Americans was attributable primarily to heart disease, tuberculosis, influenza, nephritis, cancer, pellagra, and malaria (Johnson, 1966). In fact, New York Life's and Equitable's actuaries predicted that blacks would be extinct by the year 2000 as the result of the extremely high mortality rate (Haller, 1971). Congress responded to the high death rates among African-Americans with the passage of the Freedmen's legislation, which opened universities, hospitals, soup kitchens, and clinics in the South (Blassingame, 1973; Morais, 1967).

African-American Health During the 20th Century

The late 1800s and the beginning of the 1900s were characterized by significant migration of African-Americans from rural to urban areas. The Great Migration to northern urban areas, which began in 1915, was associated with pull factors in the North, such as employment opportunities, and push factors from the South, including a depressed demand for labor, low wages, floods, segregation, discrimination, lynching, and poor educational opportunities (Woodson, 1969).

African-Americans continued to suffer from serious health problems despite migration to urban areas. In 1900, the death rate among Atlanta's blacks exceeded that among Atlanta's white population by 69 percent. Of the 431 black children born that year, 194 (45 percent) died before their first birthday, generally as the result of treatable and preventable childhood diseases (Galishoff, 1985). The president of Atlanta's Chamber of Commerce explained the high death rate as a function of blacks' "unhygienic and unsanitary modes of living, and their established susceptibility to disease" (Galishoff, 1985: 26). Atlanta's decision to improve the sanitary conditions in black neighborhoods was directly attributable to concern for the white populace:

> Because from that segregated district Negro nurses would still emerge from diseased homes, to come into our homes and hold our children in their arms; Negro cooks would still bring bacilli from the segregated district into the homes of the poor and the rich white Atlantan; Negro chauffeurs, Negro butlers, Negro laborers would come from within the pale and scatter disease with the same old lavishness; into that district would go the clothes of white families, to be laundered in environments possibly reeking with filth and disease.
> The disease germ knows no color or race line....
> (*Atlanta Constitution*, cited in Galishoff, 1985: 29)

During the 1920s, tuberculosis was responsible for three times as many deaths among blacks as among whites in New York City. Harlem's rate of infant mortality from 1923 to 1927 was 111 per 1,000, compared to a rate of 64.5 per 1,000 for the entire city of New York (Osofsky, 1966). Such disparities, particularly in the South, have been attributed in part to the actions of the white medical community (Charatz-Litt, 1992). White physicians often refused to treat black patients. Black physicians were rendered less effective in treating patients due to their inferior medical training and their exclusion from membership in many medical associations and societies, thereby precluding them from accessing new techniques (Byrd and Clayton, 1992; Charatz-Litt, 1992; Seham, 1964).

The National Hospital Association (NHA) was organized in 1923 as a member of the National Medical Association (NMA). Although the NMA's mission emphasized the education of its black physician members, the NHA focused on equality for blacks in the Southern health care system (Charatz-Litt, 1992). It was not until the mid-1960s, however, that the American Medical Association (AMA) reaffirmed its intent to cease racially discriminatory exclusion policies and practices (Anonymous, 1965). The movement toward recognition of black physicians was due in large measure to the passage of federal legislation, such as the Civil Rights Act, requiring cessation of discriminatory and exclusionary policies and practices (Byrd and Clayton, 1992).

Blacks were often solicited as subjects of medical experiments. M. Robert Hines obtained spinal fluid from 423 sick and healthy black infants at an Atlanta hospital, apparently without the permission of the children's parents or guardians. A number of the children suffered trauma, including blood in the spinal fluid, as a result of the needle puncture (Roberts, 1925).

The Tuskegee Syphilis Study

In 1929, the United States Public Health Service (USPHS) conducted a study to examine the prevalence of syphilis among blacks and possible mechanisms for treatment. The town of Tuskegee, located in Macon County in Alabama, was found to have the highest rate of syphilis among the six counties that had been included in the study (Gill, 1932; Jones, 1981). This study, funded by the Julius Rosenwald Fund, concluded that mass treatment of syphilis would be feasible. However, funding became inadequate for the continuation of the project and the implementation of the treatment due to the economic depression that commenced in 1929 and which devastated the Fund's resources (Thomas and Quinn, 1991).

The Tuskegee syphilis study was initiated in 1932 by the USPHS to follow the natural history of untreated, latent syphilis in black males. The impetus for the study derived in part from conflict between the prevailing scientific view in the United States of the progression of syphilis in blacks and the results of a study by Brussgard in Norway. The U.S. view held that syphilis affected the neurological functioning in whites, but the cardiovascular system in blacks. Bruusgard, however, had found from his retrospective study of white men with untreated syphilis that the cardiovascular effects were common and neurological complications rare (Clark and Danbolt, 1955). However, even at the time that the Tuskegee study was initiated, there existed general consensus within the medical community that syphilis required treatment even in its latent stages, despite the toxic effects of treatment. Moore (1933: 237), a venereologist, observed:

> Though it imposes a slight though measurable risk of its own, treatment markedly diminishes the risk from syphilis. In latent syphilis . . . the probability of progression, relapse, or death is reduced from a probable 25-30 percent without treatment to about 5 percent with it; and the gravity of the relapse if it occurs, is markedly diminished.

Interest in other racial differences also provided impetus to continue with the study. Blacks were believed to possess an excessive sexual desire, a lack of morality (Hazen, 1914; Quillian, 1906), and an attraction to white women stemming from "racial instincts that are about as amenable to ethical culture as is the inherent odor of the race . . ." (Howard, 1903: 424).

The original Tuskegee study was to include black males between the ages of 25 and 60 who were infected with syphilis. The study required a physical examination, x-rays, and a spinal tap. The original design did not contemplate the provision of treatment to those enrolled in the study, despite existing consensus in the medical community regarding the necessity of treatment (Brandt, 1985). However, those recruited for the study were advised that they were ill with "bad blood," a term referring to syphilis, and would be provided with treatment. The mercurial ointment and neoarsphenamine provided to subjects as treatment were ineffective and intended to be ineffective. Similarly, the spinal tap which was administered for diagnostic purposes only was portrayed as a "special treatment" to encourage participation. A control group of healthy uninfected men was added to the study as controls in 1933, following USPHS approval to continue with the study (Brandt, 1985).

The researchers themselves noted the conditions that made this extended study possible: follow-up by a nurse who was known to the participants and who came from the community from which they were recruited; the provision to the subjects of the research burial assistance, which they might not have otherwise been able to afford; the provision of transportation to the subjects by the nurse; and government sponsorship of the "care" that the subjects believed, and had been led to believe, was being furnished to them (Rivers, Schuman, Simpson, and Olansky, 1953).

The Tuskegee study continued for 40 years, despite various events that should have signaled its termination. First, the USPHS had begun to administer penicillin to some syphilitic patients in various treatment clinics (Mahoney et al., 1944). By at least 1945, it was clear in the professional literature that syphilis infections would respond to treatment with penicillin, even in cases that had been resistant to treatment with bismuth subsalicylate and mapharsen, a then-standard treatment (Noojin, Callaway, and Flowet, 1945). Yet, subjects of the Tuskegee study were not only not offered penicillin treatment, but were also prevented from receiving care when they sought it out (Thomas and Quinn, 1991). Second, a series of articles had been published in professional journals indicating that the subjects were suffering to a much greater degree than the controls with increased morbidity and a reduction in life expectancy (Deibert and Bruyere, 1946; Heller and Bruyere, 1946; Pesare, Bauer, and Gleeson, 1950; Vonderlehr, Clark, Wenger, and Heller, 1936). Yet, defenders of the study asserted as late as 1974 that there was inadequate basis for treatment with either penicillin or other regimens during the course of the study and that it was the "*shibboleth* of informed consent . . . born in court decisions in California (1957) and Kansas (1960)" that provoked the furor over the study (Kampmeier, 1974: 1352). Third, the Nuremberg Code of 1949, discussed below, enunciated standards to guide medical research that should have caused the researchers involved with the Tuskegee study to question the propriety of the study's continuation, if not its initiation. However, the acceptance of the Code among Western nations failed to have such an impact.

It was not until 1972 that the then-existing Department of Health, Education, and Welfare convened an advisory panel in response to the criticism triggered by media coverage of the experiment (Brandt, 1985). The report of the committee focused on the failure to provide penicillin treatment and the failure to obtain informed consent. According to Brandt (1985), this emphasis obscured the historical facts regarding the availability of drug treatment for syphilis prior to the advent of penicillin and ignored the fact that the men believed that they were receiving clinical care and did not know that they were part of an experiment (Brandt, 1985).

The Tuskegee study has had a far-reaching impact. The study has, for many blacks, become a "symbol of their mistreatment by the medical establishment, a metaphor for deceit, conspiracy, malpractice, and neglect, if not outright racial genocide" (Jones, 1992: 38). Small's (cited in Shavers-Hornaday, Lynch, Burmeister, and Torner, 1997: 33) statement reflects the legacy of slavery, combined with the Tuskegee experiment:

> Our whole relationship to whites had been that of their practicing genocidal conspiratorial behavior on us from the whole slave encounter up to the Tuskegee Study. People make it sound nice by saying the Tuskegee 'Study.' But do you know how many thousands and thousand of our people died of syphilis because of that?

As a consequence, educational programs designed to combat HIV in black communities have been met with distrust and a belief that AIDS and AIDS prevention and care represent forms of racial genocide (Jones, 1992; Klonoff and Landrine, 1999; Thomas and Quinn, 1991).

THE NAZI EXPERIMENTS

It was the events of World War II that ultimately provided the impetus for the development of formal international standards to guide experimentation involving human participants: the Nuremberg Code, discussed below. As with the Tuskegee study, it is important to understand the social and historical context which permitted these events to occur.

Eugenics and Racial Hygiene

Discussion regarding the perceived superiority/inferiority of various groups dates back to at least 1727, when various noblemen of France argued that they represented the descendants of a superior race, unlike the lower estates of French society, who they claimed had descended from subjugated Celtic Gauls (de Boulainvilliers, 1727, cited in Proctor, 1988). Even during the Age of Enlightenment, when philosophers readily espoused the ideals of liberty, equality, and fraternity, it was argued that Negroes and women were inferior to men and did not merit an award of the same rights (Farr, 1986). de Gobineau argued in his *Essay on the Inequality of the Human Races* (1853-1855) that race constituted the primary force underlying the great transformations in history and that racial history constituted a science. His depiction of race as a science represented a major departure from many earlier arguments that had framed racial and ethnic prejudice in terms of religious doctrine (Proctor, 1988).

Darwin's publication of the *Origin of the Species* in 1859 became a major turning point in the development of biological determinism and scientific racism (Proctor, 1988). Darwin's theory of natural selection was used by American social Darwinists to demonstrate the moral superiority of industrial capitalism, while German social Darwinists argued that the theory supported the need for state intervention to halt the degeneration of the human species (Hofstadter, 1944; Graham, 1977). Ploetz (1895) argued against medical care for the weak, claiming that their survival would result in the reproduction of others who would never survive without medical intervention, while Haycraft (1895), a British social Darwinist, referred to diseases such as tuberculosis and leprosy as "our racial friends," claiming that they attacked only those with a weaker constitution.

Ploetz, in particular, made significant efforts to advance his point of view and its impact on others. He founded in 1904 the *Journal of Racial and Social Biology* [*Archiv für Rassen- und Gesellschaftsbiologie*] and in 1905 founded, with other colleagues, the Society for Racial Hygiene [*Gesellschaft für Rassenhygiene*]. By 1943, the Society had changed its name, had increased its membership dramatically, had over 40 local branches in Germany, and had numerous foreign affiliates.

Racial hygiene became a theme in the professional literature. Siemens warned that the poor were reproducing at a higher rate than were the rich. Bluhm asserted that medical intervention was permitting women to survive childbirth who would have otherwise dies and was ultimately creating a dependency of women on medical aid. von Gruber opined that Germany was experiencing a decline in its birth rate in comparison with other European nations due to voluntary contraception and sterility caused by venereal disease.

Initially, racial hygiene appeared not to be racist. Ploetz, for instance, specifically commented on the cultural and scientific achievements of various Jews and rejected the concept of a "pure" race. However, Ploetz also argued that whites constituted the superior race and recognized the concept of Nordic supremacy as an integral component of the Society for Racial Hygiene. And, while the early years of the racial hygiene movement encompassed both liberals and reactionaries, conservative nationalist forces controlled most of the important centers of German racial hygiene by the end of World War I. Lehmann, one of Germany's leading medical publishers of the time, furthered the dissemination of these views through the publication of various treatises, commentaries, and journals. In 1934, he received the Nazi's Golden Medal of Honor (*Goldene Ehrenzeichen*) in recognition of his efforts.

By 1930, the ties between the racial hygienists, the Nordic supremacists, and the Nazi party had become quite close, so much so that by the mid-1930s, it became difficult to distinguish the rhetoric of the racial hygienists from the political platform of the Nazi party. Although adherents to nonracist racial hygienics persisted, they were strongly criticized for their views. The Nazi party was able to capitalize on longstanding anti-Semitic sentiment in the general populace for its own purposes (Weyers, 1998).

The Lamarckian theory of inheritance, which espoused the idea that acquired characteristics could be inherited, was widely discussed during this time as well and was contrasted with Mendel's scientific experiments with plants that demonstrated the principles of assortment and segregation. Opinions on the question of biological inheritance tended to divide within the German biomedical community along political lines (Proctor, 1988). Hereditary theory became a key issue throughout the Nazi regime, and the Party's views of heredity were incorporated into its dogma. For instance, its *Handbook for the Hitler Youth* asserted, after reviewing various scientific experiments, that:

> What we need to learn from these experiments is the following: Environmental influences have never been known to bring about the formation of a new race. That is one more reason for our belief that a Jew remains a Jew, in Germany or in any other country. He can never change his race, even by centuries of residence among other people. (Brennecke, 1937: 45-47, cited in Proctor, 1988:38)

The establishment in 1927 of the Kaiser Wilhelm Institute for Anthropology, Human Genetics and Eugenics helped to institutionalize racial hygiene in Germany. The mission of that center was to provide knowledge that would aid in the prevention of the "physical and mental degradation of the German people" (Proctor, 1988: 39). The research agenda was to include investigations into the effects of alcohol and venereal disease, into the heritability of various disorders such as crime and feeble-mindedness, and an analysis of demographic trends and genealogies (Proctor, 1988). Twin studies became an integral component of this government-supported research. The SS physician Josef Mengele, who had been a graduate student under the Institute's Dr. Ottmar Freiherr von Verschuer, supplied the Institute with "scientific materials" acquired at the Auschwitz concentration camp. Proctor (1988:45) has characterized the interplay of the Nazis and the German medical community:

> German biomedical scientists thus participated in a broad program of racial research. The Nazis found biology and medicine a suitable language in which to articulate their goals; scientists found the Nazis willing to support many of their endeavors. Furthermore, racial hygiene was not "imposed on" the German medical community; physicians eagerly embraced the racial ideal and the racial state.

Gerhard Wagner (cited in Lenz, 1933: 1572, cited in Proctor, 1988: 45], a leader in the German medical profession, boasted:

> Knowledge of racial hygiene and genetics has become, by a purely scientific path, the knowledge of an extraordinary number of German doctors. It has influenced to a substantial degree the basic world view of the State, and indeed may even be said to embody the very foundations of the present state [*Statstsraison*].

By the mid-1930s, the concept of racial hygiene was intimately associated with National Socialism. Fritz Lenz was one of the foremost leaders of racial hygiene throughout the 1920s and 1930s. Lenz became Germany's first professor of racial hygiene in 1923. During the first few years of the Second World War, he offered suggestions on how to racially restructure the occupied East. Together with his colleagues Eugen Fischer and Erwin Baur, he co-authored a genetics textbook that remained one of Germany's most renowned treatises on the subject for over 20 years. In 1972, he was honored by a neo-Nazi journal as the grandfather of Germany's racial hygiene movement (Proctor, 1988). Lenz argued not only that the races differed from each other in their levels of intelligence, but that the differences between the sexes were so great that men and women were to be considered different organisms. While Lenz argued that racial science demonstrated particular characteristics of Jews, such as precociousness, wittiness, empathy, and skill in the control and exploitation of other men rather than of nature, he also asserted that his views were not anti-Semitic:

> No race can be regarded as either "higher" or "lower" than another, because all such estimates of value imply the application of some standard of value other than that of race per se. We cannot say that the earth stands higher or lower than the planet Mars, or that the earth is at the same level as Mars, because the concepts "high," "low," and "level" are coined with reference to the earth itself. (Baur, Fischer, and Lenz, 1927:692, cited in Proctor, 1988: 56)

The Recruitment of Physicians to the Nazi Cause

Hitler appealed directly to the German medical establishment to aid him in his campaign of racial hygiene. In response to his plea, a group of 49 German doctors in 1929 formed the Nationalist Socialist Physicians' League (Nationalsozialistischer Deutscher Arztebund) in order to coordinate the Nazi medical policy and "to purify the German medical community of Jewish Bolshevism" (Proctor, 1988: 65). Even before Hitler's ascendancy to power, a total of 2,786 physicians, representing 6 percent of the entire medical profession, had joined this League. In fact, from 1925 to 1944, physicians joined the Nazi Party at approximately three times the rate of the general German population (Kater, 1983). The greatest support came from younger physicians under the age of 40 (Proctor, 1988).

Proctor has offered several possibilities to explain this penchant to join the Party: (1) the conservative nature of the medical profession in general, (2) the overcrowding of the medical profession and the scarcity of positions, which could be lessened with the expulsion and exclusion of Jews from the profession, and (3) the possibility that physicians would have increased power and prestige under the new regime. As of 1926, many physicians in Germany were earning slightly more than the average industrial worker. In 1929, almost one-half of the physicians in German were earning a lower wage than that which was required for minimum survival, and by 1932, almost three-quarters of German physicians were in this economic state (Hilberg, 1990, cited in Weyers, 1998). It was estimated that by 1936, almost 5,000 German medical school graduates would be unemployed (Schoeps & Schlör, 1995, cited in Weyers, 1998). The Nazi Party promised that the "misery of the rising generation of German physicians . . . will immediately be solved when, in the future Third Reich, fellow Germans will be treated only by physicians of German descent" (Johnson, 1987: 343). On March 23, 1933, Gerhard Wagner appealed to his fellow non-Jewish physicians in the Nazi Party newspaper, *Völkischer Beobachter*:

> There is hardly any profession more important for the greatness and future of the nation than the medical; no other has been so strictly organized for decades. And yet, no other is so Jewified and so hopelessly involved in unsociable thinking. Jewish lecturers dominate the chairs of medicine, disgrace medical science, and have saturated generations of young physicians with a mechanistic attitude. Jewish "colleagues" installed themselves in the managing boards of professional organizations; they debased the medical conception of honor, and undermined race-specific ethics and morality. Jewish "colleagues" gained control over our professional policy; thanks to them, a bargaining mentality and unworthy commercial attitude has increasingly established itself in our ranks. And the end of this dreadful development is the economic bankruptcy, the loss of our esteem with the people, and the continuously decreasing influence in state and administration . . . . Honor and sense of duty demand from us to put an end to this untenable situation. Therefore, we call upon all German physicians: Clean up the boards of our organizations, sweep away all who do not want to understand the signs of our time, make our profession German again in leadership and spirit. (Johnson, 1987: 342)

German medical history was essentially reconstructed to enhance the contributions of non-Jewish physicians and researchers and disown the contributions of those who were Jewish or were related to Jews, including even Nobel Prize winners and internationally respected physicians and surgeons (Weyers, 1998). The Reich's 1938 Order for Academic Promotion commanded:

> It is not possible to prohibit entirely the quotation of Jewish authors in doctoral theses. However, Jewish authors may be quoted only rarely and briefly, even if no other literature is available. In individual cases, control of compliance to this rule must be the duty of the faculty. Principally, there are no reservations against quoting Jewish authors if this is done for the purpose of disproving or fighting their ideas. At any rate, the fact that Jewish literature has been employed in any work must be disclosed, and bibliography in regard to Jewish authors must be restricted to material that is deemed absolutely necessary. (Bleker and Jachertz, 1989: 14. Cited in Weyers, 1998: 58-59).

Jewish physicians were progressively subjected to further restrictions. They were prohibited from treating non-Jews and, by 1935, the radical party press called for the death penalty for all Jewish physicians who had failed to heed this prohibition. On July 15, 1938, Hitler decreed that the licenses of all Jewish physicians be withdrawn and the medical practices of all Jewish physicians ceased (Weyer, 1998). It is estimated that, beginning in 1933, in response to continuing social, economic, and political persecution, as well as outright torture by members of the brown-uniformed storm troopers of the Nazi Party, approximately 5 percent of all Jewish physicians living in Germany during the Nazi period committed suicide (Johnson, 1987).

The "Cleansing" of the Aryan "Body"

The Office of Racial Policy was established on May 1, 1934. This office is credited with the development of the Nazi government's principal racial programs, including the Sterilization Law, the secret sterilization of the offspring of black French occupation troops and native Germans (*Rheinlandbastarde*), and the Nuremberg Laws (Proctor, 1988). Sterilization was adopted as a legal means of "improving the race," and was modeled after similar policies in Switzerland, Denmark, Norway, Sweden, Finland, Estonia, Iceland, Cuba, Czechoslovakia, Yugoslavia, Lithuania, Latvia, Hungary, Turkey, and the United States. It was the laws of the United States and their implementation, however, that most impressed the drafters of the Sterilization Law. By 1939, more than 30,000 people in 29 American states had been sterilized. In addition, many southern states in the United States maintained antimiscegenation laws prohibiting the intermarriage of the races.

The Sterilization Law provided that doctors were to register every case of genetic illness known to them from among the patients. They were to do so without making known to the patient that this information was being disclosed to the genetic health court, which would determine the patient's reproductive future. "Genetic illness" referred to illnesses such as feeble-mindedness, schizophrenia, and alcoholism. Sterilization was effectuated by vasectomy for men and tubal ligation for women. Because tubal ligation often required extended hospital stays, injections of carbon dioxide into fallopian tube tissue and radiation were later utilized as alternative techniques. However, these procedures not infrequently resulted in serious complications, including lung embolisms and death (Proctor, 1988).

The 1935 Nuremberg Laws consisted of three measures designed to "cleanse" the German population: (1) the Reich Citizenship Law, which distinguished between residents and citizens, who were defined as those "of German or related blood who through their behavior make it evident that they are willing and able faithfully to serve the German people and nation" (Proctor, 1988: 130); (2) the Law for the Protection of German Blood and German Honor, which prohibited marriage and sexual relations between Jews and non-Jews; and (3) the Law for the Protection of the Genetic Health of the German People, which prohibited the marriage of those with venereal disease, feeble-mindedness, epilepsy, or other genetic infirmities and required that couples intending to marry submit to a physical examination in order to assure that no "racial damage" would ensue from their union. "Sexual traffic" between Germans and Jews was deemed to be "racial pollution" (Proctor, 1988: 132). What constituted "Jewishness" for the purpose of these laws was carefully delineated based on an individual's ancestry.

The Nazi physicians were charged with the task of uncovering the key to "Jewish knowing" and "Jewish psychology," which was responsible for the ills of then-current medical practice: too complicated, too incomprehensible, too analytical (Böttcher, 1935, cited in Proctor, 1988: 166).

The sterilization campaign slowed after the initiation of the Second World War in 1939, due in part to the replacement of sterilization with euthanasia and, in part, to disagreements within the leadership as to the value and the parameters of sterilization (Proctor, 1988). It has been estimated that a total of 350,000 to 400,000 individuals were involuntarily sterilized by the end of this campaign (Proctor, 1992). Euthanasia was justified with the argument that, just as the healthy must sacrifice their lives during a war, so, too, should the sick. Accordingly, a campaign was waged to sacrifice "life not worth living" (Proctor, 1988: 182). Euthanasia was effectuated by injecting morphine, by tablet, and by gassing with cyanide, carbon monoxide, or chemical warfare agents. Some physicians chose to withhold care from the children and patients in their institutions, and argued that their deaths were the result of "natural causes," albeit starvation and exposure to cold. Relatives of these patients were informed that their children and other family members had died of such complications as appendicitis, brain edema, or other complication. Proctor (1988:193) has emphasized that physicians were never ordered to kill their psychiatric patients or their institutionalized children or elderly patients. Rather, they were empowered to do so and often performed this task on their own initiative.

Ultimately, anti-Semitism was medicalized through references to Jews as parasites or cancers in the body of the German people. Nazi physicians claimed that Jews suffered from a disproportionately high rate of certain metabolic and mental disorders. This racial degeneracy was attributed to the hybrid character of the Jewish race. The confinement of Jews to ghettos in Nazi-occupied territories was justified as a hygienic measure to prevent epidemics. Not surprisingly, the confinement of large numbers of individuals in relatively small areas with limited access to food and supplies resulted in outbreaks of typhoid fever, typhus, and tuberculosis, as well as in starvation and physical abuse. These epidemics provided the Nazi physicians with a rationale for the complete isolation and extermination of the Jewish population.

Regulations Governing Human Experimentation

During this period, physicians were not without guidelines and directives to assist them in discerning the appropriateness of conduct related to human experimentation. One instance of experimentation involving human subjects had, in particular, prompted the formulation of various directives and guidelines.

In 1898, Albert Neisser, a professor of dermatology and venereology at the University of Breslau, injected cell free serum from patients with syphilis into other patients, many of whom were prostitutes, without their knowledge or consent. This experiment represented an effort to develop "vaccination" against syphilis. Neisser attributed the resulting syphilis infections in the prostitutes to their work, rather than to his injections. The public prosecutor investigated the case. The Royal Disciplinary Court fined Neisser and ruled that he should have obtained the consent of the patients to give them the injections. In addition, the Scientific Office of Medical Health, which had been commissioned to investigate the case by the minister for religious, educational, and medical affairs, concluded that a physician who suspected that the injection of a serum into a patient might result in an infection must both inform the individual and obtain his or her consent prior to conducting the experiment (Vollman and Winau, 1996).

A directive issued by the Prussian Minister of Religious, Educational and Medical Affairs on December 29, 1900 stated:

> I. I wish to point out to the directors of clinics, polyclinics and similar establishments that medical interventions for purposes other than diagnosis, therapy and immunization are absolutely prohibited, even though all other legal and ethical requirements for performing such interventions are fulfilled if:
> 1. The person in question is a minor or is not fully competent on other grounds;
> 2. The person concerned has not declared unequivocally that he consents to the intervention;
> 3. The declaration has not been made on the basis of a proper explanation of the adverse consequences that may result from the intervention.
>
> II. In addition, I prescribe that:
> 1. Interventions of this nature may be performed only by the director of the institution himself or with his special authorization;
> 2. In every intervention of this nature, an entry must be made in the medical case-record book, certifying that the requirements laid down in Items 1-3 of Section I and Item 1 of section II have been fulfilled, specifying details of the case;
>
> III. This directive shall not apply to medical interventions intended for the purpose of diagnosis, therapy, or immunization. (*Centralblatt der gesamten Unterrichtsverwaitung in Preussen*, 1901: 188-189)

A later document entitled "Regulations on New Therapy and Human Experimentation" and available to German physicians reflected provisions similar to what would later be adopted as the Nuremberg Code:

> The Reich Health Council [Reichsgesundheitsrat] has set great store on ensuring that all physicians receive information with regard to the Following guidelines. The Council has agreed that all physicians in open or closed health care institutions should sign a commitment to these guidelines while entering their employment....
>
> 1. In order that medical science may continue to advance, the initiation of appropriate cases of therapy involving new and as

yet insufficiently tested means and procedures cannot be avoided. Similarly, scientific experimentation involving human subjects cannot be completely excluded as such, as this would hinder or even prevent progress in the diagnosis, treatment, and prevention of diseases.

The freedom to be granted to the physician accordingly shall be weighed against his special duty to remain aware at all times of his major responsibility for the life and health of any person on whom he undertakes innovative therapy or perform an experiment.

2. For the purposes of these Guidelines, "innovative therapy" means interventions and treatment methods that involve humans and serve a therapeutic purpose, in other words, that are carried out in a particular, individual case in order to diagnose, treat, or prevent a disease or suffering or to eliminate a physical defect, although their effects and consequences cannot be sufficiently evaluated on the basis of existing experience.

3. For the purposes of these Guidelines, "scientific experimentation" means interventions and treatment methods that involve humans and are undertaken for research purposes without serving a therapeutic purpose in an individual case, and whose effects and consequences cannot be sufficiently evaluated on the basis of existing experience.

4. Any innovative therapy must be justified and performed in accordance with the principles of medical ethics and the rules of medical practice and theory.

In all cases, the question of whether any adverse effects that may occur are proportionate to the anticipated benefits shall be examined and accessed.

Innovative therapy may be carried out only if it has been tested in advance in animal trials (where these are possible).

5. Innovative therapy may be carried out only after the subject or his legal representative has unambiguously consented to the procedure in light of relevant information provided in advance.

Where consent is refused, innovative therapy may be initiated only if it constitutes an urgent procedure to preserve life or prevent serious damage to health and prior consent could not be obtained under the circumstances.

6. The question of whether to use innovative therapy must be examined with particular care where the subject is a child or a person under 18 years of age.

7. Exploitation of social hardship in order to undertake innovative therapy is incompatible with the principles of medical ethics.

8. Extreme caution shall be exercised in connection with

innovative therapy involving live microorgamisms, especially live pathogens. Such therapy shall be considered permissible only if the procedure can be assumed to be relatively safe and similar benefits are unlikely to be achieved under the circumstances by any other method.

9. In clinics, polyclinics, hospitals, or other treatment and care establishments, innovative therapy may be carried out only by the physician in charge or by another physician acting in accordance with his express instructions and subject to his complete responsibility.

10. A report shall be made in respect of any innovative therapy, indicating the purpose of the procedure, the justification for it, and the manner in which it is carried out. In particular, the report shall include a statement that the subject, or where appropriate, his legal representative has been provided in advance with relevant information and has given his consent.

    Where therapy has been carried out without consent, under the conditions referred to in the second paragraph of section 5, the statement shall give full details of these conditions.

11. The results of any innovative therapy may be published only in a manner whereby the patient's dignity and the dictates of humanity are fully respected.

12. Section 4-11 of these Guidelines shall be applicable, *mutatis mutandis*, to scientific experimentation (cf. Section 3).

    The following additional requirement shall apply to such experimentation:
    (a) Experimentation shall be prohibited in all cases where consent has not been given;
    (b) Experimentation involving human subjects shall be avoided if it can be replaced by animal studies. Experimentation involving human subjects may be carried out only after all data that can be collected by means of those biological methods (laboratory testing and animal studies) that are available to medical science for purposes of clarification and confirmation of the validity of the experiment have been obtained. Under these circumstances, motiveless and unplanned experimentation involving human subjects shall obviously be avoided;
    (c) Experimentation involving children or young persons under 18 years of age shall be prohibited if it in any way endangers the child or young person;
    (d) Experimentation involving dying subjects is incompatible with the principles of medical ethics and shall therefore be prohibited.

13. While physicians and, more particularly, those in charge of hospital establishments may thus be expected to be guided by a strong sense of responsibility toward their patients, they

should at the same time not be denied the satisfying responsibility [*verantwortungsfreudigkeit*] of seeking new ways to protect or treat patients or alleviate or remedy their suffering where they are convinced, in the light of their medical experience, that known methods are likely to fail.
14. Academic training courses should take every suitable opportunity to stress the physician's special duties when carrying out a new form of therapy or a scientific experiment, as well as when publishing his results. (*Reichgesundheitblatt* 11, No. 10, March 1931)

The Medical Experiments

Prisoners of the Nazi concentration camps, notably Dachau, Auschwitz, Buchenwald, and Sachsenhausen, were subjected to numerous "experiments" designed to gain knowledge that would ultimately benefit the German military. At least 26 different types of experiments were carried out (Caplan, 1992). Experiments included the ingestion of seawater, subjection to extremes of cold temperature or high or low pressure, bone and limb transplants without medical necessity for such procedures, and injection with infectious bacteria to assess the effectiveness of new antibacterial drugs (Proctor, 1988). Mengele conducted experiments with twins whereby he attempted to create a Siamese twin by connecting the blood vessels and organs of two twins. Ultimately, the children died of infection (Mozes-Kor, 1992). Prisoners were deliberately infected with malaria to test the effectiveness of various antimalarial agents. Wounds were deliberately inflicted on prisoners and then infected with mustard gas to provide an opportunity to assess the effectiveness of various treatments for mustard gas-induced burns. Women, in particular, were subjected to injuries that were designed to resemble bullet wounds and battle-caused infections, so as to test various potential treatments. Still other prisoners were deliberately infected with typhus, some in order to serve as research subjects in the evaluation of an antityphus vaccine, and others in order to maintain a steady supply of the virus (Taylor, 1946). Both men and women were subjected to radiation and subsequent surgery to evaluate the effectiveness of the x-ray as a means of castration-sterilization. There were frequent complications and deaths (Lifton, 1986). The experimental radiation treatment of men's fungal infections due to shared razors often resulted in the impairment of salivary and tear-duct functions and paralysis of the face and eyes. Blond-haired, brown-eyed children were subjected to injections of methylene blue in an attempt to permanently change their eye color; these injections resulted, instead, in blindness and death (Lifton, 1986).

Prisoners forced into these experiments were not told what was going to happen to them and were not given an opportunity to either consent or to refuse to consent. There was no attempt to minimize any of the risks that they faced (Mozes-Kor, 1992). As Lifton (1986:14) so eloquently expressed, "[a]t the heart of the Nazi enterprise [was] the destruction of the boundary between healing and killing."

Significant questions arising from the Nazi experiments and physicians' involvement in them continue to demand analysis and answers. These include the ethical and scientific issues raised in conjunction with the current use of the data obtained through these experiments (Berger, 1992; Freedman, 1992; Katz and Pozos, 1992; Pozos, 1992); the possible existence of similarities between what we call euthanasia today and the Nazis' use of the term and the ethical implications of such similarities if they do, indeed, exist (Macklin, 1992); and whether the Nazi experiments represented "an isolated aberration in the history of medical experimentation" (Katz, 1992: 235) or, instead, constitute the most extreme example of the ubiquitous and relatively milder abuse of human beings in the name of medical research (Katz, 1992).

THE COLD WAR EXPERIMENTS

Literally thousands of human radiation experiments were conducted during the Cold War using, in general, subjects who were poor, sick, and powerless (Welsome, 1999). Radiation experiments on soldiers, including the repeated insertion of radium rods into their nostrils and the administration of irradiated foods to conscientious objectors, began in the 1940s or early 1950s and continued until 1962 (Josefson, 1997). This section describes only several of the many experiments conducted on civilians. Although some of these experiments were initiated prior to the Nuremberg Code, many were continued and/or commenced after its promulgation.

The Cold War: How It Began and What It Meant

The beginnings of what has become known as The Cold War have been traced to negotiations conducted between the British, Russian, and American heads of state in Yalta from February 4 to February 11, 1945, in an effort to obtain a Soviet commitment to enter the Second World War and to "rectify" boundaries in eastern Europe (Winks, 1964). Winks has asserted that, inappropriately, short-term military considerations took precedence in these negotiations over long-range diplomatic goals. Lynd (1976: 16) has attributed much of the responsibility for the deterioration of U.S.-Russian relations on former President Roosevelt's reluctance to address realistically issues of power that required resolution:

> Why did the cold war start? The most fundamental answer might be: Because for the first time the challenge of authoritarian socialism to democratic capitalism was backed by sufficient power to be an ever-present political and military threat. It is a far more complicated and potent challenge than that represented by Germany in 1914 or Japan in 1941; it is the kind of challenge associated with the breakup of empires and the transformation of whole societies rather than with the ordinary jostling of diplomatic intercourse.

Russia honored its Yalta agreement to enter the war through its invasion in Manchuria. Soon thereafter, Roosevelt announced the unconditional surrender of the Japanese forces, followed soon by the surrender of the Nazi forces. In rapid succession, the Charter of the United Nations had been signed and a conference was held in Potsdam, Germany to decide the postwar fate of Europe. Agreements entered into by the Russians at Yalta were not honored at Potsdam; the Soviet government wanted to establish a ring of satellite countries around it, which the United States, France, Great Britain, and China refused to permit but could not physically prevent. Churchill later remarked, coining the term, the "iron curtain,":

> A shadow has fallen upon the scenes so lately lighted by the Allied victory. Nobody knows what Soviet Russia and its Communist international organization intends to do in the immediate future, or what are the limits, if any, to their expansive and proselytizing tendencies . . . . [A]ll these famous [European] cities and the populations around them lie in the Soviet sphere and are subject in one form or another, not only to Soviet influence but to a very high and increasing measure of control from Moscow....In front of the iron curtain which lies across Europe are other causes for anxiety.... (Quoted in Winks, 1964: 27)

Montgomery (1997: xv) described the impact of the Cold War on the academic environment:

> The Cold War, and in particular the commitment of both the United States and the Soviet Union to rapid development of massive stockpiles of weapons with which they could exterminate each other, fueled the rapid expansion of business and academic activity, which continued for thirty years.

The politics of the Cold War in the United States promoted the targeting of individuals as "atomic spies," the political cleansing of various industries and professions (Montgomery, 1997), the initiation of research agendas based on political ideology and the pressures of American foreign policy, and an influx of government monies into academic institutions and industry to carry out these research agenda (Lewontin, 1997). From 1949 and 1952, the epitome of the McCarthy era in the United States, dissidents in the Soviet Union were being imprisoned or executed on charges of "rootless cosmopolitanism" and the "worship of things foreign" (Montgomery, 1997).

The Vanderbilt Nutrition Study

The Rockefeller Foundation funded, in part, a study conducted by Vanderbilt University to determine how a woman's diet and nutrition affected her pregnancy and delivery, as well as the condition of her infant. Women were enrolled into the radioactive iron experiment without either their knowledge or their consent. Pregnant women were told that they were to receive a "cocktail." As researchers later admitted, this "cocktail" had no known therapeutic value. The women who were administered the radioactive iron later experienced rashes, bruises, anemia, a loss of teeth and hair, and cancer. Four of the children who had been exposed to prenatal radiation developed fatal malignancies; no cancers were found among the children of unexposed mothers (Hagstrom, Glasser, Brill, and Heyssel, 1969).

The Fernald State School Experiments

Established in 1848, the Fernald State School was the first permanent school for "feeble-minded" children. Later, the school also housed boys from abusive, poor and unstable families (Welsome, 1999). However, the idealistic principles upon which it was originally founded were eroded by a growing sentiment that demanded that society be protected from the mentally retarded, rather than protecting the mentally retarded. Fernald, a psychiatrist for whom the school was eventually renamed, characterized the retarded as

> a parasitic, predatory class, never capable of self-support or of managing their own affairs...Feeble-minded women are almost invariably immoral and if at large usually become carriers of venereal disease or give birth to children who are as defective as themselves....Every feeble-minded person, especially the high-grade imbecile, is a potential criminal....(Clarke and Clarke, 1966: 16).

One of the parents of a Fernald School resident described in her testimony to the Massachusetts Task Force on Human Subject Research (1994: 38) the conditions that existed there:

> The first Sunday of each month was the only visiting time allowed. When I arrived, I was not allowed beyond the foyer....On visiting, I occasionally got a glimpse of the day room. It was a large, bare room with a cement-like floor. In the middle of the room there was a circular grating where urine and feces were hosed down. Needless to say, the little girls wore no panties. This room had no chairs, the children sat or laid down on the cold flooring. Most of the children laid down on the floor because they were given Valium two and three times a day....

The floor was also populated by roaches, red ants, and mice.
    Between 1946 and 1953, the Massachusetts Institute of Technology (MIT) conducted an experiment in which 74 boys at Fernald received trace amounts of radioactive calcium or iron in their oatmeal. The experiment was designed to assess the extent to which children were deprived of important minerals due to the presence of chemicals (phylates) that combined with calcium and iron to form insoluble compounds. Such experiments had been discouraged in normal children. (Welsome, 1999). The boys participating in these experiments were rewarded with admittance into the Science Club. A later investigation revealed that the parents of the children were not informed about the radioactive component of the study, but were told only that their children would receive a special breakfast that included

calcium and would be rewarded for their participation with membership in the Science Club (Moreno, 1999; Welsome, 1999). A letter sent to the parents by the Fernald School superintendent in 1949 described the study as follows:

> We are very much interested in the various aspects of nutrition, particularly how the body absorbs various cereals, iron and vitamins. We are considering the selection of a group of our brighter patients . . . to receive a special diet rich in the above-mentioned substances for a period of time. It will be necessary to make some blood tests at stated intervals, similar to those to which our patients are already accustomed, and which will cause no discomfort or change in their physical condition other than possibly improvement. (Farrell, 1949)

Total Body Irradiation at the University of Cincinnati

Beginning in 1960, and continuing until 1972, Eugene Saenger conducted total body irradiation studies with patients suffering from various forms of cancer. The experiments, funded by the Defense Atomic Support Agency, were designed to utilize increasing dosages of radiation, commencing with 100 rads and increasing the exposure up to 600 rads. Saenger had observed that fatalities could begin to occur at the exposure level of 200 rads.

The purpose of these experiments was clearly nontherapeutic. In fact, the program was designed specifically to benefit the military:

> This program is designed to obtain new information regarding the metabolic, physiologic, immunologic, hematologic, and biochemical effects of Total Body Radiation and Partial Body Radiation in human beings. It will then be possible to understand better the influence of radiation on combat effectiveness of troops and to develop more suitable methods of diagnosis, prognosis, prophylaxis and treatment of radiation injuries. (Saenger, 1966:1)

Patients receiving "treatment" did so in a sitting position with their heads tilted forward and their legs raised, in order to mimic the posture of a soldier assuming a defensive position (Saenger, 1966). The vast majority of the participants were elderly, poorly educated, African American, and dependent on charity for their medical care. Many were also cognitively impaired, with an average IQ of 89, compared to a "normal" IQ of 100 (Egilman, Wallace, Stubbs, and Mora-Corrasco, 1998). Saenger and colleagues (1973) acknowledged that the total body irradiation had contributed to the deaths of at least 8 patients. Numerous patients died soon after receiving the administration of radiation (Welsome, 1999).

Members of the Universty's Faculty Committee on research, which reviewed the original and revised protocols, voiced serious concerns relating to the ethical legitimacy of the experiments and the consent forms to be presented to the prospective participants (Egilman, Wallace, Stubbs, and Mora-Corrasco, 1998).

The consent forms used at the initiation of these experiments do not reveal what was told to the patients about the potential risks of their participation but stated:

> The nature and purpose of this therapy, possible alternative methods of treatment, the risks involved, the possibility of complication, and prognosis have been fully explained to me. The special study and research nature of this treatment has been discussed with me and understood by me. (Egilman, Wallace, Stubbs, and Mora-Corrasco, 1998: 77)

Although later versions of the consent form disclosed the potential for infection and bleeding, they did not disclose the risk of death. Notably, the National Institutes of Health in 1969 and again in 1973 refused to fund the experiments on ethical grounds (Welsome, 1999).

Scientists at Oak Ridge were approached by the Atomic Energy Commission in 1966 to conduct similar experiments. The offer was refused due toe ethical concerns:

> The suggestion is made that we should treat carcinoma of the breast, gastrointestinal tract, and urogenital tract by total body irradiation. These groups of patients have been very carefully considered for such therapy, and we are very hesitant to treat them because we believe there is so little chance of benefit to make it questionable ethically to treat them. Lesions that require moderate or high doses of local therapy for benefit, or that are actually resistant (gastrointestinal tract) are not helped enough by total body irradiation to justify the bone marrow depression that is induced. (Hearing on TBI Program, 1981).

The Aftermath

Ultimately, an advisory commission held hearings to ascertain whether the radiation experiments had had a clear medical or scientific purpose and whether they had complied with the ethical standards that prevailed at the time (Welsome, 1999). The Committee identified six basic ethical principles that it found relevant to its mission:

1. One ought not to treat people as mere means to the ends of others.
2. One ought not to deceive others.
3. One ought not to inflict harm or risk of harm.
4. One ought to promote welfare and prevent harm.
5. One ought to treat people fairly and with equal respect.
6. One ought to respect the self-determination of others.

(Advisory Committee on Human Radiation Experiments, 1996: 405). In addressing the issue of whether current ethical principles could be applied to past conduct, the Committee found that

> some principles are so basic that we ordinarily assume, with good reason, that they are applicable to the past as well as the present (and will be applicable in the future as well). We regard these principles as basic because any minimally acceptable ethical standpoint must include them.
>
> While basic ethical principles do not change, interpretations and applications of basic ethical principles as they are expressed in more specific rules of conduct do evolve over time through processes of cultural change. Recognizing that specific moral rules do change has implications for how we judge the past. . . [T]he concept of informed consent has undergone refinement and development . . . . (Advisory Committee on Human Radiation Experiments, 1996: 405-406)

Testimony indicated that at least some of the experimenters were familiar with the principles enunciated in the Nuremberg Code, as well as with the code of ethics espoused by the American Medical Association (AMA). Committee members found that patients and society in general, at the time that the radiation experiments had been conducted, ceded extensive decision making authority to physicians and that the medical profession at that time did not generally believe that informed consent was a prerequisite to the enrollment of a patient in research. One committee member characterized the Nuremberg Code as aspirational in nature and commented that experimenters had complied with ethical codes more in form than in substance (Welsome, 1999). The committee's ultimate determination that most of the experiments were harmless provoked severe criticism:

> Beyond the question of harm, beyond the evil of duplicity, the most unfortunate casualty of the Cold War radiation agenda was the simple capacity of individuals to make informed decisions about their own bodies. Unfortunately, the committee does not seem to lend the principle of self-determination the same value it accords some of the others in its list of moral precepts. Rather, it seems to focus on risks to patients. The panel admits that the nonconsensual use of humans in nontherapeutic experiments is always an affront, but it says "As the burden on the patient-subject decreases, so too did the seriousness of the wrong." That construction lets the government off too easily, for it does not assign blame based upon the essential nature of the action itself— the use of an innocent person as a test animal—but rather, fosters a retrospective opinion that allows less-bad outcomes to ameliorate the action's inherent wrong. The committee's recommendation that some of those experimented upon without consent deserve only apologies is informed by this belief. (Editorial, 1995: 18)

The Committee was critical, however, of experiments conducted on individuals for whom there could not possibly have been any therapeutic benefit of those experiments. This formed the basis of a criticism not only of the individual investigators involved in these efforts, but also of the medical profession itself:

> The historical silence of the medical profession with respect to nontherapeutic experiments was perhaps based on the rationale that those who are ill and perhaps dying may be used in experiments because they will not be harmed even though they will not benefit. But this rationale overlooks both the principle that people never should be used as mere means to the ends of others and the principle of respect for self-determination; it may also provide insufficient protection against harm and inadequately represent the best interests of the patient, given the position of conflict of interest in which the physician-researchers may find themselves. (Advisory Committee on Human Radiation Experiments, 1996: 405-406)

Lawsuits were ultimately filed by the families of many of the patients who had received the radioactive "treatments." Many of the universities and companies who initiated the experiments either settled the claims against them, or paid damages in accordance with resulting judgments. Quaker Oats and the Massachusetts Institute of Technology entered into a settlement in 1997 with approximately 30 of the Fernald School alumni, in the amount of $1.85 million (Moreno, 1999). A number of lawsuits were dismissed. One judge was blunt in her criticism of the experiments that had taken place:

> The allegations in this case indicate that the government of the United States, aided by the officials of the City of Cincinnati, treated at least eighty-seven of its citizens as though they were laboratory animals. If the Constitution has not clearly established a right under which the plaintiffs may attempt to prove their case, then a gaping hole in that document has been exposed. *The subject of experimentation who has not volunteered is merely an object* (Beckwith, 1995: 58) (emphasis added).

THE PRISON EXPERIMENTS

The Holmesburg Prison Experiments

Despite the enunciation of the Nuremberg Code's principles governing medical research involving humans, and the later promulgation of the Helsinki Declarations (see below), the American medical establishment failed to adopt these principles as their own. Indeed, research sponsored by pharmaceutical companies, academic institutions, and governmental entities increasingly relied on marginalized members of society as subjects for their research. Senator Kennedy observed in 1973, after the promulgation and adoption of both the Nuremberg Code and the Helsinki

Declaration, that "Those who have borne the principal brunt of research—whether it is drugs or even experimental surgery—have been the most disadvantaged people within our society; have been the institutionalized, the poor, and minority members."

Holmesburg Prison was the largest of Philadelphia's county jails (Hornblum, 1998). Medical research had begun in the county's penal system in the 1950s, under the direction of Dr. Albert Kligman, a dermatologist by training. Kligman's research record prior to its initiation at Holmesburg was questionable ethically, but remained generally unchallenged. For instance, he had previously conducted an experiment to test various fungistatic preparations by deliberately infecting retarded children with the target disease (Kligman and Anderson, 1951). His later experimental use of x-rays from 1956 to 1957 to treat fungus infections of the nail in retarded children and prisoners received the financial support of the USPHS (Hornblum, 1998).

One of the most common experiments conducted on prisoners was known as the "Patch Test." This involved the exposure of the inmate's skin to various untested skin lotions, creams, and moisturizers for a period of 30 days. Periodic exposure of the tested areas to a sunlamp often resulted in burns and blisters. The resulting patchwork design on the inmates' skin became a distinctive characteristic of men who had served time at Holmesburg. Later experiments involved the removal of thumbnails to see how fingers reacted to abuse, the inoculation of skin with herpes simplex and herpes zoster (Goldschmidt and Kligman, 1958), the experimental infection of inmates with ringworm (Strauss and Kligman, 1957), the inoculation of inmates' skin with cutaneous moniliasis (Maibach and Kligman, 1962), the infection of inmates with bacteria staphylococcus aureus (Singh, Marples, and Kligman, 1971) and candida albicans (Rebora, Marples, and Kligman, 1973), and the exposure of prisoners to phototoxic drugs (Kligman and Briet, 1968) and long ultraviolet rays (Willis and Kligman, 1969). Participation in the least desirable of these experiments was reserved for the black prisoners (Hornblum, 1998). Indeed, experimentation with prisoners as subjects appears to have been a commonplace occurrence throughout the prison system. Heller, a prison psychiatrist who worked at Holmesburg during Kligman's tenure there, had enumerated in a co-authored journal article the types of experiments being conducted in prisons: the testing of tranquilizers, analgesics, and antibiotics for dosage and toxicity for various pharmaceutical companies; a study of toothpaste and mouthwash conducted for Johnson & Johnson; a study of the absorbency and wound adhesion properties of various dressings for Johnson & Johnson; a study of antiseptic lotion for Johnson & Johnson; a study of an antiperspirant preparation for Helena Rubenstein; and a "napkin absorbency study" for DuPont, in which female prisoners were paid for each used napkin that was frozen, saved, and returned to DuPont (Heller, 1967).

Participation in such experiments constituted one of the few mechanisms available to inmates to earn money, which could be used to purchase personal hygiene supplies or to post bail. The sums earned were not inconsequential. For instance, in 1959 alone, the inmates earned a total of $73,253 by volunteering to take pills and use creams. Additionally, prisoner participation in these experiments provided prison administrators with an additional means of control; unruly prisoners could be threatened with the termination of their participation and the consequent loss of their sole source of income. It was not until years later, during an investigation into the occurrence of sexual assaults in the Philadelphia prison system, that it was revealed that the economic power gained by some of the inmates through participation in these research experiments conducted by Kligman at the University of Pennsylvania was used to coerce sexual favors from other inmates. Although investigating officials were told that an inmate was not permitted to earn more than $1,200 per year, there were instances of prisoners earning more than $1,700 in less than a year (Hornblum, 1998).

Eventually, the medical experiments at Holmesburg were ceased. The American Civil Liberties Union, concerned about the circumstances under which inmates' participation was solicited, proposed the following guidelines for the recruitment of prisoners into medical research:

1. All consents obtained for the purposes of any form of experimentation must be informed consents.
2. It is the responsibility of the researcher to make sure that the prospective volunteer is in the proper physical and (when relevant) mental condition to undertake the experiment.
3. Waiver forms or exculpatory language in the consent document must be banned.
4. Researchers must be required to carry insurance providing total coverage for subjects adversely affected by the experiment, and for compensation of the family or next of kin in case of death.
5. Prisoner-volunteers must be paid at a scale commensurate with what the researcher would offer "free world" volunteers as compensation.
6. Prisoners should not be promised reduced sentences or favorable consideration for parole in return for participation in a clinical experiment.
7. Assurances must be made and enforced that the experiment will be carried out in a manner that does not necessarily threaten the lives or safety of the prison-volunteers.
8. Subject to requirement number 6, no report or records of the prisoners' participation in the experiments should be released to anyone by the researcher or by prison authorities without the signed consent of the prisoners.
9. A supervisory committee, independent of prison authorities, the researcher, and the sponsors of the research, must be established to review and oversee all experimentation conducted in prisons.
10. The sponsors of prison research (drug companies, foundations, or whoever) must pay reasonable sums for the privilege of having access to the inmate population for research purposes.
11. Prison authorities should immediately undertake to provide greater opportunities for prisoners to work and earn money (Rudovsky, 1973).

These proposed guidelines were tendered in 1973 and the Nuremberg Code had been promulgated in 1949, the first Helsinki Declaration in 1964. It is worth noting, however, that at least as early as 1967, researchers questioned whether prisoners could ever make a free choice regarding their participation in medical experiments (Heller, 1967).

Malaria, Leukemia, and Pellagra

Holmesburg Prison was not, however, alone among prisons in its reliance on inmates for scientific experimentation. Over 400 inmates at Stateville Penitentiary in Illinois participated in a two-year long experiment intended to discover a cure for malaria (Hornblum, 1998). Participating inmates were required to sign a consent form which absolved the investigators and prison authorities of all legal liability for injuries that might arise in connection with their participation. The waiver contained the following language:

> I...hereby declare that I have read and clearly understood the above notice, as testified by my signature hereon, and I hereby apply to the University of Chicago, which is at present engaged on malarial research at the orders of the Government, for participation in the investigations of the life-cycle of the malarial parasite. I hereby accept all risks connected with the experiment and on behalf of my heirs and my personal and legal representatives I hereby absolve from such liability the University of Chicago and all the technicians and assistants taking part in the above mentioned investigations. I similarly absolve the Government of the State o fIllinois, the Director of the Department of Public Security of the State of Illinois, the warden of the State Penitentiary of Joliet-Stateville and all employees of the above institutions and Departments, from all responsibility, as well as from all claims and proceedings or Equity pleas, for any injury or malady, fatal or otherwise, which may ensue from these experiments.
>
> I hereby certify that this offer is made voluntarily and without compulsion. I have been instructed that if my offer is accepted I shall be entitled to remuneration amounting to [xx] dollars payable as provided in the above Notice....(Pappworth, 1990:62)

Prisoners in Atlanta's federal penitentiary were also recruited into malaria experiments (George, 1946). The goal of the Atlanta prison experiments was "to control malaria, by chemotherapy (including chemoprophylaxis), in the Armed Forces of the United States with the least possible delay" (George, 1946: 16). The goal was to be effectuated through a four-part strategy designed to

> (1) Fully educate anyone concerned in the use of *known* antimalarial drugs for the suppression and therapy of malaria infection. (2) Evolve new drugs which would prevent, or control malaria in man—drugs that could be used safely by man over long periods of time. (3) Discover new drugs which would destroy the malaria parasite and obviate periodic malaria relapses even after evident "cure" through the use of known amtimalarial drugs such as quinine, atabrine, and plasmochin. (4) Originate drugs as effective as quinine or atabrine while at the same time occasioning less toxic reaction than is experienced with the use of these drugs. (George, 1946: 16)

A total of 130 men participated in the experiment. Although none died, all experienced the symptoms of malaria, including fevers as high as 106 degrees, chills, nausea, headaches, and backaches. A journalist reporting on the conclusion of the Atlanta Malaria Project boasted:

> After the last man has had his physical examination and been certified as *cured*, and I n good health, Atlanta's Malaria Project will then become another shining light in the galaxy of wartime achievements at Atlanta. The project will be aligned with the $30,000,000 worth of war goods produced by Atlanta's industries, the 2,250 pints of whole blood donated to the Armed Forces during thee war, the $207,000 worth of Defense, War, and Victory Bonds purchased, and the $10,000 donated to the Red Cross. And, while it is true that Atlanta's Malaria project was but one spoke in the huge wheel of medical research going on to defeat this dreadful disease, the malaria volunteers, like the GIs, accomplished the mission assigned to them. (George, 1946: 43)

Other prison-based experiments involved a California prisoner in Sing Sing was who transfused with blood from a leukemia patient Anon., 1949a, b, c, d, e), inmates of the federal prison at Chillicothe, Ohio who were enrolled in an investigation of an oral polio vaccine (Anon., 1956b), and prisoners of the Ohio prison system who were recruited into a study undertaken by Ohio State University and the Sloan-Kettering Institute for Cancer Research, in which they received injections of live cancer cells in an effort to "discover the secret of how healthy human bodies fight the invasion of malignant cells" (Anon., 1956a). Experiments sponsored by the Central Intelligence Agency (CIA) during the 1960s tested numerous psychotropic drugs on prisoners, including those at Holmesburg, who were not told what they were ingesting (Hornblum, 1998). Inmates in prisons in Utah, Colorado, Oklahoma, California, Pennsylvania and Illinois participated in numerous irradiation experiments sponsored by the Atomic Energy Commission. Inmates were not informed that they were receiving such substances, and were not advised of the potential adverse effects (Advisory Committee on Human Radiation Experiments, 1995; Lee, 1994).

In fact, one earlier experiment using prisoners was so well known that one of the Nazi physicians referred to it at the time of the Nuremberg Trial. Dr. Joseph Goldberger had been a public health official. Convinced that pellagra was due to a lack of protein in the diet, rather than poor sanitation and a variety of other postulated causes, he embarked on an experiment to test his theory. He convinced then-governor of Mississippi Earl Brewer to allow him to conduct an experiment with a dozen inmates of Rankin Farm prison. He would induce the onset of pellagra through the gradual modification of their diet to one consisting solely of sweet potatoes, grits, rice, and cornbread. The volunteers, who had been promised a pardon for their crimes in exchange for their participation, soon began to suffer from dizziness, lethargy, pain, and skin lesions. One prisoner characterized his participation as having been through "a thousand hells" and another claimed that he would prefer a "lifetime of hard labor" rather than endure such a "hellish experiment" (Etheridge, 1972: 7).

THE DEVELOPMENT OF DIETHYLSTILBESTEROL (DES)

Diethylstilbesterol, or DES, is a synthetic estrogen. It was developed in London in 1938 and was prescribed during the period from 1945 to 1971 to prevent spontaneous abortions. Initial studies of the drug's efficacy were conducted at Harvard University in the late 1940s. Although the investigators did not have a control group against which to compare the results, they concluded from their findings that the use of DES would result in a healthier maternal environment (Weitzner and Hirsch, 1981). A later study at Tulane University, however, found that the women who had been treated with DES had more miscarriages and premature births in comparison with the mothers who had not used DES.

Pregnant women at the Lying-in Hospital of the University of Chicago were randomized to a clinical trial, with one-half of the women receiving DES and the other half receiving placebo. The researchers found that women receiving DES were more likely to have miscarriages and small babies than those who had been given placebo. None of the women were informed that they were participating in research, none were told what drug they were taking, and none were asked for consent to participate. In 1951, the Food and Drug Administration concluded that DES was safe for use in pregnancy (Mascaro, 1991). The drug continued to be used for 20 years. By 1971, it was estimated that 1.5 million babies had been exposed to DES. In 1971 alone, approximately 30,000 had been exposed (Weitzner and Hirsch, 1981).

Researchers reported in 1971 that between 1966 and 1971, 7 cases of clear-cell adenocarcinoma had been found in teenage girls (Herbst, Ulfelder, and Poskanzer, 1971). The one element that appeared to be common to these rare cases of cancer was the ingestion of DES by the girls' mothers during their pregnancies. Injuries to female babies of DES mothers have included vaginal and cervical dysplasia, adenosis, uterine structural abnormalities, infertility, menstrual irregularities, fetal death and premature birth, and breast and reproductive-tract cancers (Weitzner and Hirsch, 1981). Injuries to male babies have also been reported, including penile bleeding, testicular masses, hypoplastic testes, and sterility (NIH, 1992; Weitzner and Hirsch, 1981).

Numerous lawsuits have been filed against the University of Chicago for the injuries alleged to have resulted from this experiment. In at least two cases, the lawsuits have been settled prior to trial.

THE WILLOWBROOK HEPATITIS EXPERIMENTS

Willowbrook State School was a state-funded, -licensed, and –operated institution for the severely mentally retarded, located in New York State. Krugman (1986: 158-159) described the conditions that prevailed at Willowbrook, which had been built to house several thousand residents and, instead, held 6,000:

> Of the 6,000 [residents], there were 77% who were severely and profoundly retarded, 60% who were not toilet trained, 39% who were not ambulatory, 30% who had seizures, and 64% who were incapable of feeding themselves....The annual attack rate of hepatitis with jaundice 25 per 1,000 among children and 40 per 1,000 among adults....[U]nder the conditions existing in the institution, most newly admitted children would contract hepatitis. This empiric impression was confirmed in the 1970s when newly developed serologic tests revealed that >90% of the residents of the institution had hepatitis A and B markers of the past infection.

Beginning in the 1950s, Saul Krugman of New York University conducted studies of hepatitis virus at Willowbrook State School, utilizing the mentally retarded institutionalized there as his research subjects. Krugman reasoned that the majority of children admitted to the state school would ultimately develop hepatitis because of the poor conditions there. Consequently, children were deliberately infected with hepatitis virus in order to study the natural history of the disease. The experiment was criticized on ethical grounds in the medical literature for: (1) the failure to protect the children against hepatitis through the use of gamma globulin injections, which had been found to be efficacious; (2) the subtle pressure exerted on parents to give their permission for their children to participate in such experiments, by expediting the admission of their children into the school; and (3) the failure to inform fully the parents of these children of the risks that attended their children's participation in the experiment (Beecher, 1970).

THE TEAROOM TRADE

Humphreys, a doctoral candidate in sociology at the University of Washington, conducted research on the characteristics of men who sought quick and impersonal sexual gratification from other men. At the time, the public and the police often held to stereotypes regarding homosexuality.

Humphreys served as the "watchqueen," at various "tearooms," stationing himself near the men participating in sex so that he could keep watch and cough at the approach of a police car or stranger. He eventually revealed his identity and motivation for serving in this role. Many of the better educated clientele agreed to speak with him about their practices. Humphries was concerned about the potential for bias resulting from a sample that was weighted heavily to those of higher socioeconomic status. He secretly followed some of the other men and recorded their license plate numbers. He then matched these numbers with the records of the department of Motor Vehicles. Approximately one year later, Humphreys visited these men, and, representing himself as a health service interviewer, obtained extensive data regarding their marital status and sexual behaviors.

As a result of Humphreys' findings, it was learned that many of the men seeking anonymous sexual gratification with other men identified themselves neither as homosexual nor bisexual. Over one-half of them were married and living with their wives. Many of the men or their wives were Catholic and, often due to family planning concerns, sexual relations were infrequent. Only approximately 14 percent of the individuals interviewed were primarily interested in homosexual relations (Humphreys, 1970).

Although Humphreys' research was later praised by many social scientists and gay organizations, it raised significant ethical concerns. First, Humphreys' data could have been accessed by law enforcement agencies through subpoenas, thereby compromising the men's privacy and the confidentiality of the data collected. (Subpoenas are discussed in chapter 5.) Second, the men who were observed, and those who were later followed, did not know of the nature of the research, or even that it was research, at the time of their initial contact with Humphreys.

Humphreys relied on the research strategy known as "deception" to obtain his data. This practice, still common in some social science disciplines in particular, raises serious questions regarding an individual's right to make informed choices regarding the disclosure of information related to him- or herself and the ultimate impact on the perceived integrity of scientists and science itself. Deception as a research strategy is discussed in more depth in conjunction with study design, in chapter 3.

THE DEVELOPMENT OF INTERNATIONAL CODES AND GUIDELINES

The Nuremberg Code and Declaration of Helsinki

It was ultimately the events during World War II that provided the impetus for the development of formal international standards to guide experimentation involving human participants: the Nuremberg Code. The Nuremberg Code (1949) enumerates ten basic principles that are deemed to be universally applicable to research involving human subjects:

1. The prospective participant's voluntary consent is essential.
2. The results to be obtained from the experiment must be beneficial to society and those results cannot be obtained through any other means.
3. The study must have as its foundation the results of animal experiments and a knowledge of the natural history of the disease so that the anticipated results justify the conduct of the experiment.
4. All unnecessary physical and mental injury or suffering must be avoided during the course of the experiment.
5. No experiment should be conducted if it is believed that death or disabling injury will occur, except where the research physicians also serve as research participants.
6. The degree of risk should not exceed the humanitarian importance of the problem being addressed.
7. Adequate facilities and preparations should be used to protect the participant against death or injury.
8. Only scientifically qualified persons should conduct the experiment.
9. The participant has a right to end his or her participation at any time if he or she reaches a point where continuation seems impossible.
10. The scientist in charge must be prepared to end the experiment if there is probable cause to believe that continuing the experiment will likely

result in the injury, disability, or death of the research participant (World Medical Association, 1991b; Annas and Grodin, 1992).

The Nuremberg Code has been subject to a great deal of criticism, particularly for its failure to distinguish between therapeutic clinical research and clinical research on healthy participants and to provide a review mechanism for researchers' actions (Perley, Fluss, Bankowski, and Simon, 1992). These deficiencies and the resulting discussions ultimately led to the formulation and adoption of the Declaration of Helsinki.

The Declaration of Helsinki was initially adopted by the World Medical Association in 1964. Unlike the Nuremberg Code, the Declaration (1) allows participation in research through the permission of a surrogate, where the actual participant is legally or physically unable to consent to his or her participation and (2) distinguished between clinical research combined with professional care and nontherapeutic clinical research. Later revisions in 1975, 1983, and 1989 further emphasized the need for an individual's voluntary informed consent to participate in research (World Medical Association, 1991a; Perley et al., 1992; Christakis and Panner, 1991). Unlike later documents, the Helsinki Declaration emphasizes the role of physicians in conducting research and does not impose a requirement of review of research protocols by an independent body prior to the initiation of the research.

Although the Nuremberg Code has often been called "universal," it is clear that this is not the case. For instance, the expert witnesses who appeared at the Nuremberg Trial indicated that the standards which they were reciting for the conduct of experimentation involving human subjects were already in place in the United States. However, as we saw from the discussion above, this was far from accurate. Additionally, these experts utilized the writings of Western civilization as the basis for their assertions, without regard to other traditions (Grodin, 1992). Universalists contend that the universal recognition and/or adoption of specified standards is critical in order to prevent the exploitation of more vulnerable and less sophisticated populations and societies. Pluralists and relativists argue in response that the imposition of Western standards on a society constitutes yet another form of exploitation. The concept of informed consent, for instance, reflects the concept of an individual as a free-standing being, whose decisions are uninhibited by weight considerations of others. This is a uniquely Western concept of the individual and, as such, fails to provide essential guidance to the conduct of research in societies and cultures where concepts of personhood may differ greatly (see Loue, Okello, and Kawuma, 1996). The universality of informed consent as visualized by the various international codes and guidelines is discussed further in chapter 4.

The International Covenant on Civil and Political Rights

The International Covenant on Civil and Political Rights (ICCPR) was drafted by an international committee following World War II. Although it was entered into force in 1976, the United States did not ratify it until 1992. As of 1997, it had been ratified by more than 115 countries, representing more than two-thirds of the world's population (Rosenthal, 1997).

Article 7 of ICCPR provides that:

> No one shall be subjected to torture or to cruel, inhuman or degrading treatment or punishment. In particular, no one shall be subjected without his free consent to medical or scientific experimentation.

In essence, the clause links nonconsensual experimentation to torture and inhuman or degrading treatment, one of the most fundamental prohibitions of international law.

The Article itself, however, fails to provide any guidance with respect to experimentation. That inadequacy is addressed, in part, by General Comment 20, adopted by the Human Rights Committee in 1992:

> Article 7 expressly prohibits medical or scientific experimentation without the free consent of the person concerned . . . The Committee also observes that special protection in regard to such experiments is necessary in the case of persons not capable of giving valid consent, and in particular those under any form of detention or imprisonment.

Unlike the Nuremberg Code and the Helsinki Declaration, discussed above, and the CIOMS documents, discussed below, the ICCPR is binding on those countries that have ratified it; the countries agree to enforce the ICCPR through their own legal systems (Vennell, 1995). And, although ratifying countries must report to the United Nations Human Rights Committee the mechanisms that they have adopted to effectuate the provisions of ICCPR, there is no international mechanism for the enforcement of the ICCPR provisions (Rosenthal, 1997). In the United States, the treaty is non-self-executing, so that a private right f action does not exist (Stewart, 1993).

CIOMS and WHO International Guidelines

International documents were later developed to provide further guidance in the conduct of research involving humans and to address some of the deficiencies of the Nuremberg Code and the Helsinki Declaration. The *International Guidelines for Ethical Review of Epidemiological Studies* were compiled by the Council for International Organizations of Medical Sciences (CIOMS, 1991) specifically to aid researchers in the field of epidemiology to resolve moral ambiguities that may arise during the course of their research. The *International Ethical Guidelines for Biomedical Research Involving Human Subjects* (1993), prepared by CIOMS and the World Health Organization (WHO), set forth a statement of general ethical principles, 15 guidelines, and relevant commentary reflecting both the majority and minority points of view. These two documents, in particular, reflect the potential for heterogeneity across cultures in which the research is to be conducted, and diversity of discipline among the investigators carrying out the research studies. As such, they reflect broadly stated principles and offer various mechanisms for the application of those principles, which may differ greatly among cultures. The WHO's *Guidelines for Good Clinical Practice for Trials on Pharmaceutical Products* (1995) provides additional guidelines for the conduct of clinical trials. Specific provisions of these documents are discussed throughout the text, beginning with chapter 3.

THE DEVELOPMENT OF UNITED STATES GUIDELINES AND REGULATIONS

The development of formal regulations and guidelines for the conduct of research involving humans developed along a different time line than did the international codes. One might think, in view of the seemingly extensive research conducted through the 1970s in the United States without participant knowledge or consent that, apart from the international codes and guidelines, nothing existed to guide the conduct of the health researchers. Such was not the case, although ethical thought relating to human experimentation was not developed in the 1940s to the extent indicated by the American expert witnesses at the Nuremberg Trial.

As early as 1871, a New York court cautioned physicians against unnecessary experimentation with their patients:

> [W]hen the case is one as to which a system of treatment has been followed for a long time, there should be no departure from it unless the surgeon who does it is prepared to take the risk of establishing, by his success, the propriety and safety of his experiment.
> This rule protects the community against reckless experiments, while it admits the adoption of new remedies and modes of treatment only when their benefits have been demonstrated, or when, from the necessity of the case, the surgeon or physician must be left to the exercise of his own skill and experience. (*Carpenter v. Blake*, 1871: 524)

And, in 1935, a Michigan court stated:

> We recognize the fact that, if the general practice of medicine and surgery is to progress, there must be a certain amount of experimentation carried on; but such experiments must be done with the knowledge and consent of the patient or those responsible for him, and must not vary too radically from the accepted method of practice. (*Fortner v. Koch*, 1935).

In 1953, with the opening of the NIH Clinical Center, guidelines were formulated to govern informed consent in research (Frankel, 1975). These guidelines, however, applied only to volunteers who did not have any physical illnesses. Written consent would be required where there was a possibility of an unusual hazard (Sessoms, 1963). Research funded by NIH, but conducted elsewhere in the country, was not subject to regulation.

In 1960, the NIH funded a study to review procedures governing research in major medical centers in the United States. Of the 52 centers responding to the questionnaire, 9 indicated that they had established procedures to guide their investigators in the design and conduct of research, 5 indicated that they were planning to formulate guidelines, and 22 had committees that reviewed protocols but did not address issues related to recruitment (Curran, 1970).

Congress, however, enacted the Drug Amendments of 1962 in response to public concerns stemming from the births in Europe of large numbers of severely deformed infants to women who had ingested thalidomide during their pregnancies. These amendments required that individuals participating in clinical trials of new drugs be informed of the investigative purpose of the trial and that the investigators obtain informed consent as a prerequisite to participation (Kelsey, 1963).

In 1966, the Public Health Service issued a statement requiring that recipients of grant monies from the NIH

> provide prior review of the judgment of the principal investigator or program director by a committee of his institutional associates. This review should assure an independent determination: (1) of the rights and welfare of the individual or individuals involved, (2) of the appropriateness of the methods used to secure informed consent, and (3) of the risks and potential medical benefits of the investigation (Frankel, 1975: 52).

At least in part, the impetus for the development and implementation of this requirement for independent ethical review came from the disclosure of the Tuskegee study and of various other research experiments that had been conducted under questionable circumstances (*Barrett v. Hoffman*, 1981; Beecher, 1966; Jones, 1981; Katz, 1972). These independent review committees, or institutional review boards, are discussed in detail in chapter 3.

Also in response to disclosures of questionable research, Congress passed the National Research Act in 1974. This legislation established the National Commission for the Protection of Human Subjects of Biomedical and Behavioral Research, which was charged with the responsibility of identifying the ethical principles that govern research involving human subjects and the formulation of appropriate guidelines.

In fulfillment of this charge, the Commission published in 1978 the Belmont Report. The Commission defined a principle as "a general judgment that serves as a basic justification for the many particular prescriptions for and evaluation of human actions " (1978: 4). The Commission identified three basic principles, which would then provide the basis for the development of rules and norms: respect for persons, beneficence, and justice. Respect for persons was defined to incorporate the obligations to treat individuals as autonomous agents and to provide additional protections to individuals with diminished autonomy. Beneficence was said to encompass two rules: that of doing no harm, and that of maximizing benefits and minimizing harms. Justice, as explained by the Commission, related to the equitable sharing of benefits and burdens of research (National Commission for the Protection of Human Subjects of Biomedical and Behavioral Research, 1978). In formulating these principles, the Commission did not rely on a specific approach to bioethics, although each of the principles may appear to reflect particular approaches to the analysis of ethical issues (Abram and Wolf, 1984). These various approaches are discussed in chapter 2.

CHAPTER SUMMARY

The experiments discussed in this chapter have numerous elements in common. They relied on classes of individuals tat were particularly vulnerable in one or more ways: children, who lacked an understanding of what was to occur; poor and marginalized citizens; seriously ill individuals; and prisoners, whose ability to truly volunteer was questionable in view of the degree of control exercised over their lives. The experiments were often conducted without the knowledge or consent of those conscripted into such service. Individuals were not apprised of either the procedures that they were to undergo or the risks that they would face. And, ultimately, many of those individuals were physically, mentally, and/or emotionally harmed.

We see that various international and national guidelines and laws have been developed in response to such events. These documents were formulated in order to safeguard research participants from harm. Chapter 2 will discuss various approaches to ethical analysis from which the principles enumerated in these documents may drive.

EXERCISE

Alan Cantwell, Jr., M.D. has stated: "To those perceptive enough to discern it, the mass deaths of homosexuals from AIDS was similar to the mass deaths of Jews in the Holocaust."

1. Compare and contrast HIV-related human research with the human research conducted with the Jews during the Holocaust. Include in your response a discussion of the following items as they relate to both gay-AIDS research and Jewish-Holocaust era research:

    a. the nature of the research conducted;

    b. the procedures used to enroll participants in the research;

    c. the historical and social contexts in which the research occurred and is occurring.

2. Assume that you are conducting a study which is examining the relationship between various types of social support and actual reductions in HIV risk behaviors.

a. To what extent, if any, might the attitude/approach reflected in Cantwell's statement impact on your ability to recruit and retain participants in your research study? Support your response using historical examples.
b. If you believe that this attitude will impact on your recruitment efforts, what strategies will you utilize to mediate the impact of this view on your recruitment of participants? If you do not believe that there will be an impact, or that any particular strategies are required, state so and support your answer.

References

Abt, I. (1903). Spontaneous hemorrhages in newborn children. *Journal of the American Medical Association, 40*, 284-293.

Abram, M.B. & Wolf, S.M. (1984). Public involvement in medical ethics: A model for government action. *New England Journal of Medicine, 310*, 627-632.

Advisory Committee on Human Radiation Experiments. (1995). *Final Report of the Advisory Committee on Human Radiation Experiments*. Washington, D.C.: Author.

Advisory Committee on Human Radiation Experiments. (1996). Research ethics and the medical profession: Report of the Advisory Committee on Human Radiation Experiments. *Journal of the American Medical Association, 276*, 403-409.

Annas, G.J. & Grodin, M.A. (eds.). (1992). *The Nazi Doctors and the Nuremberg Code: Human Rights in Experimentation*. London: Oxford University Press.

Anonymous. (1949a). Blood exchanger named. *New York Times*, June 23.

Anonymous. (1949b). Convict joins own blood stream to that of girl dying of cancer. *New York Times*, June 4, at 1.

Anonymous. (1949c). Convict's blood gift fails to save girl. *New York Times*, June 8, at 1.

Anonymous. (1949d). Pardoned lifer returns to son. *New York Times*, December 24.

Anonymous. (1949e). Sing Sing lifer freed by Dewey. *New York Times*, December 23, at 1.

Anonymous. (1956a). Cancer by the Needle, *Newsweek*, June 4, 67.

Anonymous. (1956b). New live polio vaccine, taken orally, to get mass test. *New York Times*, October 7.

Anonymous. (1965). AMA reaffirms effort to end discrimination. *Journal of the National Medical Association, 57*, 314.

Appelbaum, P.S., Lidz, C.W., & Meisel, A. (1991). *Informed Consent: Legal Theory and Clinical Practice*. London: Oxford University Press.

Barrett v. Hoffman, 521 F. Supp. 307 (S.D.N.Y. 1981).

Baur, E., Fischer, E., & Lenz, F. (1921). *Grundriss der menschlichen Erblichkeitslehre und Rassenhygiene*. Vol. 1, *Menschliche Erblichkeitslehre*, 3rd ed., 1927. Translated as *Human Heredity*, London, 1931. Cited in R.N. Proctor. (1988). *Racial Hygiene: Medicine Under the Nazis*. Cambridge, Massachusetts: Harvard University Press.

Beckwith, S.S. (1958). Opinion and Order, U.S. District Court for the Southern District of Ohio, Western Division (Case No. C-1-94-126), January 11.

Beecher, H.K. (1966). Ethics and clinical research. *New England Journal of Medicine, 274*, 1354-1360.

Beecher, H.K. (1970). *Research and the Individual: Human Studies*. Boston: Little, Brown and Company.

Belais, D. (1910). Vivisection animal and human. *Cosmopolitan, 50*, 267-273.

Berger, R.L. (1992). Nazi science: Comments on the validation of the Dachau human hypothermia experiments. In A.L. Caplan (Ed.), *When Medicine Went Mad: Bioethics and the Holocaust* (pp. 109-133). Totowa, New Jersey: Humana Press.

Blassingame, J. W. (1973). *Black New Orleans, 1860-1880*. Chicago, Illinois: University of Chicago Press.

Bleker J. & Jachertz, N. (Eds.). (1989). *Medizin in "Dritten Reich."* Cologne: Deutscher Ärtze-Verlag. Cited in W. Weyers. (1998). *Death of Medicine in Nazi Germany: Dermatology and Dermatopathology Under the Swastika*. A. B. Ackerman (Ed.). Philadelphia: Ardor Scribendi, Ltd.

Böttcher, A. (1935). Die Losung der Judenfrage, *Ziel und Weg, 5*, 379, cited in R.N. Proctor. (1988). *Racial Hygiene: Medicine Under the Nazis*. Cambridge, Massachusetts: Harvard University Press.

Brandt, A.M. (1985). Racism and research: The case of the Tuskegee syphilis study. In J.W. Leavitt & R.L. Numbers (Eds.), *Sickness and Health in America: Readings in the History of Medicine and Public Health* (pp. 331-343). Madison, Wisconsin: University of Wisconsin Press.

Brennecke, F. (Ed.). (1937). *Vom deutschen Volk und seinem Lebensraum, Handbuch für die Schullungsarbeit in der Hitler Jugend*. Munich. Trans. by H.L. Childs. (1938). *The Nazi Primer: Official Handbook for Schooling the Hitler Youth*. New York. Cited in R.N. Proctor. (1988). *Racial Hygiene: Medicine Under the Nazis*. Cambridge, Massachusetts: Harvard University Press.

Byrd, W.M. & Clayton, L.A. (1992). An American health dilemma: A history of blacks in the health system. *Journal of the National Medical Association, 84*, 189-200.

Caplan, A.L. (1992). How did medicine go so wrong? In A.L. Caplan (Ed.), *When Medicine Went Mad: Bioethics and the Holocaust* (pp. 53-92). Totowa, New Jersey: Humana Press.

Carpenter v. Blake, 60 Barb. 488 (N.Y. Sup. Ct. 1871).

Centralblatt der gesamten Unterrichtsverwaitung in Preussen . (1901). Trans. by Health Legislation Unit of the World Health Organization. In M.A. Grodin. (1992). Historical origins of the Nuremberg Code. In G.J. Annas & M.A. Grodin (Eds.), *The Nazi Doctors and the Nuremberg Code: Human Rights in Human Experimentation* (pp. 121-144). New York: Oxford University Press.

Charatz-Litt, C. (1992). A chronicle of racism: The effects of the white medical community on black health. *Journal of the National Medical Association, 84*, 717-725.

Christakis, N.A. & Panner, M.J. (1991). Existing international ethical guidelines for human subjects research: Some open questions. *Law, Medicine, & Health Care, 19*, 214-221.

Clarke, A.M. & Clarke, W.N. (Eds.). (1966). *Mental Deficiency: The Changing Outlook*. New York: Free Press.

Clark, E.G. & Danbolt, N. (1955). The Oslo study of the natural history of untreated syphilis. *Journal of Chronic Disease, 2*, 311-344.

Council for International Organizations of Medical Sciences (CIOMS). (1991). *International Guidelines for Ethical Review of Epidemiological Studies*. Geneva: CIOMS.

Council for International Organizations of Medical Sciences (CIOMS), World Health Organization (WHO). (1993). *International Ethical Guidelines for Biomedical Research Involving Human Subjects*. Geneva: CIOMS and WHO.

Curran, W.J. (1970). Governmental regulation of the use of human subjects in medical research: The approach of two federal agencies. In P.A. Freund (Ed.), *Experimentation with Human Subjects* (pp. 402-454). New York: Braziller.

de Boulainvilliers, H.C. (1727). *Historie de l'ancien gouvernement de la France [History of the ancient government in France]*. The Hague. Cited in R.N. Proctor. (1988). *Racial Hygiene: Medicine Under the Nazis*. Cambridge, Massachusetts: Harvard University Press.

de Gobineau, H.C. (1853-1855). *Essai sur l'inégalité des races humaines [Essay on the inequality of the human races]*. Paris. Cited in R. N. Proctor. (1988). *Racial Hygiene: Medicine Under the Nazis*. Cambridge, Massachusetts: Harvard University Press.

Deibert, A.V. & Bruyere, M.C. (1946). Untreated syphilis in the male Negro. III. Evidence of cardiovascular abnormalities and other forms of morbidity. *Journal of Venereal Disease Information, 27,* 301-314.

Douglass, F. (1968). *Narrative of the Life of Frederick Douglass, An American Slave, Written by Himself*. New York: New American Library. (Original work published 1845).

Editorial. (1995, Oct. 4). Ethical trimming on radiation. *Boston Globe*, at 18.

Egilman, D., Wallace, W., Stubbs, C., & Mora-Corrasco, F. (1998). A little too much of the Buchenwald touch? Military radiation research at the University of Cincinnati, 1960-1972. *Accountability in Research, 6*, 63-102.

Etheridge, E.W. (1972). *The Butterfly Caste*. Westport, Connecticut: Greenwood.

Farr, J. (1986). So vile and miserable an estate: The problem of slavery in John Locke's political thought, *Political Theory, 14*, 263-289.

Farrell, W.J. (1949, November 2). Consent form provided to parents. In Massachusetts Task Force on Human Subject Research. (1994*). A Report on the Use of Radioactive Materials in Human Subject Research in Residents of State-Operated facilities Within the Commonwealth of Massachusetts from 1943-1973*. Submitted to Philip Campbell, Commissioner, Commonwealth of Massachusetts, Executive Office of Health and Human Services, Department of Mental Retardation, April 1994 (Appendix B, page 19). Massachusetts: Department of Mental Retardation.

Fortner v. Koch, 261 N.W. 762 (Mich. 1935).

Frankel, M.S. (1975). The development of policy guidelines governing human experimentation in the United States: A case study of public policy-making for science and technology. *Ethics in Science & Medicine, 2*, 43-59.

Freedman, B. (1992). Moral analysis and the use of Nazi experimental results. In A.L. Caplan (Ed.), *When Medicine Went Mad: Bioethics and the Holocaust* (pp. 141-154). Totowa, New Jersey: Humana Press.

Galishoff, S. (1985). Germs know no color line: Black health and black policy in Atlanta, 1900-1918. *Journal of the History of Medicine, 40*, 22-41.

George, J. (1946). Atlanta's malaria project. *Atlantian,* 6, 14-17, 43.

Gill, D.G. (1932). Syphilis in the rural Negro: Results of a study in Alabama. *Southern Medical Journal, 25*, 985-990.

Goldschmidt, H. & Kligman, A.M. (1958). Experimental inoculation of humans with ectodermotropic viruses. *Journal of Investigative Dermatology, 31*, 175-182.

Graham, L. (1977). Science and values: The eugenics movement in Germany and Russia in the 1920s. *American Historical Review, 82*, 1122-1164.

Grodin, M.A. (1992). Historical origins of the Nuremberg Code. In G.J. Annas and M.A. Grodin (Eds.), *The Nazi Doctors and the Nuremberg Code: Human Rights in Human Experimentation* (pp. 121-144). New York: Oxford University Press.

Haller, J.S. (1971). *Outcasts from Evolution: Scientific Attitudes of Racial Inferiority.* New York: McGraw-Hill.

Hagstrom, R.M., Glasser, S.R., Brill, A.B., & Heyssel, R.M. (1969). Long term effects of radioactive iron administered during human pregnancy. *American Journal of Epidemiology, 90*, 1-10.

Hammill, S.M., Carpenter, H.C., & Cope, T.A. (1908). A comparison of the von Pirquet, Calmette and Moro tuberculin tests and their diagnostic value. *Archives of Internal Medicine, 2*, 405-447.

Haycraft, J. (1895). *Social Darwinism and Race Betterment.* London. Cited in R.N. Proctor. (1988). *Racial Hygiene: Medicine Under the Nazis.* Cambridge, Massachusetts: Harvard University Press.

Hazen, H.H. (1914). Syphilis in the American Negro. *Journal of the American Medical Association, 63*, 463-466.

Hearing on the Human Total Body Irradiation (TBI) Program at Oak Ridge Before the Subcommittee on Investigations and Oversight House Science and Technology Committee, 97th Congress, 1st Session, September 23, 1981.

Heller, J.R., Jr. & Bruyere, P.T. (1946). Untreated syphilis in the male Negro. II. Mortality during 12 years of observation. *Journal of Venereal Disease Information, 27*, 34-38.

Heller, M.S. (1967). Problems and prospects in the use of prison inmates for medical experiments. *The Prison Journal, 57*, 21-38.

Herbst, A.L., Ulfelder, H., & Poskanzer, D.C. (1971). Adenocarcinoma of the vagina. Association of maternal stilbesterol therapy with tumor appearance in young women. *New England Journal of Medicine, 284,* 878-881.

Hilberg, R. (1990). *Die Vernichtug der europäischen Juden.* Frankfurt: Fischer Taschenbuch Verlag. Cited in W. Weyers. (1998). *Death of Medicine in Nazi Germany: Dermatology and Dermatopathology Under the Swastika.* A. B. Ackerman (Ed.). Philadelphia: Ardor Scribendi, Ltd.

Hofstadter, R. (1944). *Social Darwinism in American Thought 1860-1915.* Philadelphia: University of Pennsylvania Press.

Hornblum, A.M. (1998). *Acres of Skin: Human Experiments at Holmesburg Prison.* New York: Routledge.

Howard, W.L. (1903). The Negro as a distinct ethnic factor in civilization. *Medicine (Detroit), 9,* 424.

Humphreys, L. (1970). *Tearoom Trade: Impersonal Sex in Public Places.* Chicago: Aldine.

Hurmence, B. (Ed.). (1984). *My Folks Don't Want Me to Talk About Slavery: Twenty-One Oral Histories of Former North Carolina Slaves.* Winston-Salem, North Carolina: John F. Blair.

International Covenant on Civil and Political Rights, *opened for signature* Dec. 19, 1966, 999 U.N.T.S. 171 (entered into force Mar. 23, 1976; adopted by the U.S. Sept. 8, 1992).

Jacobs, H. (1988). *Incidents in the Life of a Slave Girl.* New York: Oxford University Press.

Johnson, C.S. (1966). *Shadow of the Plantation.* Chicago, Illinois: University of Chicago Press.

Johnson, P.A. (1987). *A History of the Jews.* New York: Harper/Perennial.

Jones, J. (1981). *Bad Blood: The Tuskegee Syphilis Experiment—A Tragedy of Race and Medicine.* New York: Free Press.

Josefson, D. (1997). U.S. admits radiation experiments on 20000 veterans. *British Medical Journal, 315*, 566.

Kampmeier, R.H. (1974). Final report on the "Tuskegee syphilis study." *Southern Medical Journal, 67*, 1349-1353.

Kater, M. (1983). *The Nazi Party.* Cambridge, Massachusetts: Harvard University Press.

Katz, J. (1992). Abuse of human beings for the sake of science. In A.L. Caplan (Ed.), *When Medicine Went Mad: Bioethics and the Holocaust* (pp. 233-270). Totowa, New Jersey: Humana Press.

Katz, J. (1972). *Experimentation with Human Beings.* New York: Russell Sage.

Katz, J. & Pozos, R.S. (1992). The Dachau hypothermia study: An ethical and scientific commentary. In A.L. Caplan (Ed.), *When Medicine Went Mad: Bioethics and the Holocaust* (pp. 135-139). Totowa, New Jersey: Humana Press.

Kelsey, F.O. (1963). Patient consent provisions of the Federal Food, Drug, and Cosmetic Act. In I. Ladimer & R.W. Newman (Eds*.). Clinical Investigation in Medicine: Legal, Ethical, and Moral Aspects* (pp. 336-344). Boston: Boston University Law-Medicine Research Institute.

Kennedy, E. (1973, March 7). *Hearings on Human Experimentation*, part 3, at 841.

Kligman, A.M. & Anderson, W.W. (1951). Evaluation of current methods for the local treatment of tinea capitis. *Journal of Investigative Dermatology, 16*, 155-168.

Kligman, A.M. & Briet, R. (1968). The identification of phototoxic drugs by human assay. *Journal of Investigative Dermatology, 51*, 90-99.

Klonoff, E.A. & Landrine, H. (1999). Do blacks believe that HIV/AIDS is a government conspiracy against them? *Preventive Medicine, 28*, 451-457.

Kolchin, P. (1993). *American Slavery 1619-1877*. New York: Hill and Wang.

Krugman, S. (1986). The Willowbrook hepatitis studies revisited: Ethical aspects. *Review of Infectious Diseases, 8*, 157-162.

Lederer, S.E. (1997). *Subjected to Science: Human Experimentation in America Before the Second World War*. Baltimore: John Hopkins University Press.

Lee, A.S. & Lee, E.S. (1977). The health of slaves and the health of freedmen: A Savannah study. *Phylon, 38*, 170-180.

Lee, G. (1994, Nov. 28). The lifelong harm to radiation's human guinea pigs. *Washington Post*, National Weekly Edition, 33.

Lewontin, R.C. (1997). The Cold War and the transformation of the academy. In *The Cold War & the University: Toward an Intellectual History of the Postwar Years.* (pp. 1-34). New York: New Press.

Lifton, R.J. (1986). *The Nazi Doctors: Medical Killing and the Psychology of Genocide*. New York: Basic Books.

Loue, S., Okello, D., & Kawuma, M. (1996). Research bioethics in the Uganda context: A program summary. *Journal of Law, Medicine, & Ethics, 24*, 47-53.

Lynd, S. (1976). How the cold war began. In N.A. Graebner, (Ed.), *The Cold War: A Conflict of Ideology and Power* (2nd ed., pp. 3-17). Lexington, Massachusetts: D.C. Heath and Company.

Macklin, R. (1992). Which way down the slippery slope? Nazi medical killing and euthanasia today. In A.L. Caplan (Ed.), *When Medicine Went Mad: Bioethics and the Holocaust* (pp. 173-200). Totowa, New Jersey: Humana Press.

Mahoney, J., Arnold, R.C., Sterner, B.L., Harris, A., & Zwally, M.R. (1944). Penicillin treatment of early syphilis. II. *Journal of the American Medical Association, 126*, 63-67.

Maibach, H.I. & Kligman, A.M. (1962). The biology of experimental human cutaneous moniliasis. *Archives of Dermatology, 85*, 233-257.

Massachusetts Task Force on Human Subject Research. (1994*). A Report on the Use of Radioactive Materials in Human Subject Research in Residents of State-Operated facilities Within the Commonwealth of Massachusetts from 1943-1973*. Submitted to Philip Campbell, Commissioner, Commonwealth of Massachusetts, Executive Office of Health and Human Services, Department of Mental Retardation, April 1994. Massachusetts: Department of Mental Retardation.

Montgomery, D. (1997). Introduction: Prosperity under the shadow of the bomb. In *The Cold War & the University: Toward an Intellectual History of the Postwar Years* (pp. vii-xxxvii). New York: New Press.

Moore, J.E. (1933). *The Modern Treatment of Syphilis*. Baltimore: Charles C. Thomas.

Morais, H.M. (1967). *The History of the Negro in Medicine* (2nd ed.). New York: Publishers Co., Inc.

Moreno, J.D. (1999). *Undue Risk: Secret State Experiments on Humans*. New York: W.H. Freeman and Company.

Mozes-Kor, E. (1992). The Mengele twins and human experimentation: A personal account. In G.J. Annas & M.A. Grodin (Eds.). (1992). *The Nazi Doctors and the Nuremberg Code: Human Rights in Human Experimentation* (pp. 53-59). New York: Oxford University Press.

National Commission for the Protection of Human Subjects of Biomedical and Behavioral Research. (1978). *The Belmont Report: Ethical Principles and Guidelines for the Protection of Human Subjects of Research*. [DHEW Publication No. (OS) 78-0012]. Washington, D.C.

Noojin, R.O., Callaway, J.L., & Flower, A.H. (1945). Favorable response to penicillin therapy in a case of treatment-resistant syphilis. *North Carolina Medical Journal, January*, 34-37.

NIH (National Institutes of Health).(1991). NIH Workshop: Long-Term Effects of Exposure to Diethylstilbesterol (DES), Falls Church, Virginia, April 22-24. Sponsored by the Office of Research on Women's Health, National Cancer Institute, National Institute of Child Health and development and the National Institute of Environmental Sciences. Cited in Institute of Medicine.

(1994). *Women and Health Research*. Vol. 1. *Ethical and Legal Issues of Including Women in Clinical Studies*. Washington, D.C.: National Academy Press.

Osler, W. (1907). The evolution of the idea of experiment in medicine. *Transactions of the Congress of American Physicians and Surgeons, 7*, 7-8.

Osofsky, G. (1966). *Harlem: The Making of a Ghetto, Negro New York, 1890-1930*. New York: Harper and Row.

Pappworth, M.H. (1967). *Human Guinea Pigs*. Boston: Beacon Press.

Perley, S., Fluss, S.S., Bankowski, Z., & Simon, F. (1992). The Nuremberg Code: An International Overview. In G.J. Annas & M.A. Grodin (Ed.), *The Nazi Doctors and the Nuremberg Code: Human Rights in Human Experimentation* (pp. 149-173). New York: Oxford University Press.

Pesare, P.J., Bauer, T.J., & Gleeson, J.A. (1950). Untreated syphilis in the male Negro: Observation of abnormalities over sixteen years. *American Journal of Syphilis, Gonorrhea, and Venereal Diseases, 34*, 201-213.

Proctor, R.N. (1992). Nazi doctors, racial medicine, and human experimentation. In G.J. Annas & M.A. Grodin (Eds.). (1992). *The Nazi Doctors and the Nuremberg Code: Human Rights in Human Experimentation* (pp. 17-31). New York: Oxford University Press.

Proctor, R.N. (1988). *Racial Hygiene: Medicine Under the Nazis*. Cambridge, Massachusetts: Harvard University Press.

Quillian, D.D. (1906). Racial peculiarities: A cause of the prevalence of syphilis in Negroes. *American Journal of Dermatology & Genito-urinary Disease, 10*, 277-279.

Rebora, A., Marples, R.R., & Kligman, A.M. (1973). Experimental infection with candida albicans. *Archives of Dermatology, 108*, 69-73.

Reichgesundheitblatt 11, No. 10, (March 1931). 174-175. In International Digest of Health Legislation. (1980). 31, 408-411. Reprinted in M.A. Grodin. (1992). Historical origins of the Nuremberg Code. In G.J. Annas & M.A. Grodin (Eds.), *The Nazi Doctors and the Nuremberg Code: Human Rights in Human Experimentation* (pp. 121-144). New York: Oxford University Press.

Rivers, E., Schuman, S.H., Simpson, L., & Olansky, S. (1953). Twenty years of followup experience in a long-range medical study. *Public Health Reports, 68*, 391-395.

Roberts, M.H. (1925). The spinal fluid in the newborn. *Journal of the American Medical Association, 85*, 500-503.

Rosenthal, E. (1997). The International Covenant on Civil and Political Rights and the rights of research subjects. In A.E. Shamoo (Ed.). *Ethics in Neurobiological Research with Human Subjects: The Baltimore Conference on Ethics* (pp. 265-272). Amsterdam: Overseas Publishers Association.

Rudovsky, D. (1973, June 21). Statement of American Civil Liberties Union. Greater Philadelphia Branch Hearings on Medical Experimentation in State and County Institutions, Pennsylvania Departments of Justice & Public Welfare, Harrisburg, Pennsylvania.

Saenger, E. (1966). Metabolic changes in humans following total body irradiation. University of Cincinnati College of Medicine to the defense Atomic Support Agency, report of February 1960—April 30, 1966 (DASA-1844). Contract DA-49-146-XZ-029 [ACHRE No. DOD-042994-A-1]. Cited in Egilman, D., Wallace, W., Stubbs, C., & Mora-Corrasco, F. (1998). A little too much of the Buchenwald touch? Military radiation research at the University of Cincinnati, 1960-1972. *Accountability in Research, 6*, 63-102.

Saenger, E., Silberstein, E., Aron, B., Horwitz, B., Kereiakes, J., Bahr, G., Perry, H., & Friedman, B. (1973). Whole body and partial body radiotherapy of advanced cancer. *American Journal of Roentgenology, 117*, 670-685.

Savitt, T.L. (1978). *Medicine and Slavery: The Diseases and Health Care of Blacks in Antebellum Virginia*. Chicago, Illinois: University of Chicago Press.

Savitt, T.L. (1985). Black health on the plantation: Masters, slaves, and physicians. In J.W. Leavitt & R.L. Numbers (Eds.), *Sickness and Health in America: Readings in the History of Medicine and Public Health* (2nd ed. rev., pp. 313-330). Madison, Wisconsin: University of Wisconsin Press.

Schoeps, J.H. & Schlör, J. (1995). *Antisemitismus: Vorurteile und Mythen*. Munich: Piper. Cited in W. Weyers. (1998). *Death of Medicine in Nazi Germany: Dermatology and Dermatopathology Under the Swastika*. A. B. Ackerman (Ed.). Philadelphia: Ardor Scribendi, Ltd.

Seham, M. (1964). Discrimination against Negroes in hospitals. *New England Journal of Medicine, 271*, 940-943.

Semmes, C.E. (1996). *Racism, Health, and Post-Industrialism: A Theory of African-American Health*. Westport, Connecticut: Praeger.

Sesoms, S.M. (1963). Guiding principles in medical research involving humans, National Institutes of Health. In I. Ladimer & R.W. Newman (Eds.), *Clinical Investigation in Medicine: Legal, Ethical, and Moral Aspects* (pp. 143-147). Boston: Boston University Law-Medicine Research Institute.

Shavers-Hornaday, V.L., Lynch, C.F., Burmeister, L.F., & Torner, J.C. (1997). Why are African-Americans underrepresented in medical research studies? Impediments to participation. *Ethnicity and Health, 2,* 31-45.

Singh, G., Marples, R.R., & Kligman, A.M. (1971). Experimental aureus infection in humans. *Journal of Investigative Dermatology, 57,* 149-162.

Smith, V. (1988). Introduction. In H. Jacobs, *Incidents in the Life of a Slave Girl* (pp. xxvii-xl). New York: Oxford University Press.

Stampp, M. (1956). *The Peculiar Institution: Slavery in the Ante-Bellum South.* New York: Vintage Books.

Stewart, D.P. (1993). United States ratification of the Covenant on Civil and Political Rights: The significance of the reservations, understandings, and declarations. *DePaul Law Review, 42,* 1183-1207.

Taylor, T. (1946). Opening Statement of the Prosecution, December 9, 1946 [Nuremberg Trials]. Reprinted in G.J. Annas & M.A. Grodin (Eds.). (1992). *The Nazi Doctors and the Nuremberg Code: Human Rights in Human Experimentation* (pp. 67-93). New York: Oxford University Press.

Thomas, S.B. & Quinn, S.C. (1991).The Tuskegee syphilis study, 1932 to 1972: Implications for HIV education and AIDS risk education programs in the black community. *American Journal of Public Health, 81,* 1498-1504.

United Nations Human Rights Committee. (1992). General Comment 20(44)(art. 7), U.N. GAOR, Hum. Rts. Comm., 47th Sess., Supp. No. 40 at 194. U.N. Doc. A/47/40.

Van Evrie, J.H. (1861). *Negroes and Negro "Slavery."* New York.

Vennell, V.A.M. (1995). Medical research and treatment: Ethical standards in the international context. *Medical Law International, 2,* 1-21.

Vollman, J. & Winau, R. (1996). Informed consent in human experimentation before the Nuremberg Code. *British Medical Journal, 313,* 1445-1447.

Vonderlehr, R.A., Clark, T., Wenger, O.C., & Heller, J.R., Jr. (1936). Untreated syphilis in the male Negro. A comparative study of treated and untreated cases. *Venereal Disease Information, 17,* 260-265.

Welsome, E. (1999). *The Plutonium Files.* New York: Dial Press.

Weitzner, K. & Hirsch, H.L. (1981). Diethylstilbesterol—medicolegal chronology. *Medical Trial Technique Quarterly, 28,* 145-170.

Winks, R.W. (1964). *The Cold War: From Yalta to Cuba.* New York: Macmillan.

Woodson, C.G. (1969). *A Century of Negro Migration.* New York: Russell and Russell.

World Health Organization. (1995). *Guidelines for Good Clinical Practice for Trials on Pharmaceutical Products.* Geneva: WHO.

World Medical Association. (1991a). Declaration of Helsinki. *Law, Medicine, & Health Care, 19,* 264-265.

World Medical Association. (1991b). The Nuremberg Code. *Law, Medicine, & Health Care, 19,* 266.

2
APPROACHES TO ETHICAL ANALYSIS

This chapter reviews a number of major ethical theories, which are systematically related bodies of principles and rules. Those that have been developed or reinstated more recently are presented first and in greater length. At a less general level, but still a general level, there are ethical principles which flow from these theories and which provide the basis for rules or norms. A rule is a general statement that states that something should or should not be done because it is right or it is wrong. And, at the most specific but superficial level, there are judgments, which reflect the decision or conclusion about a specific action. Chapters 3 and 4 address the ethical principles and rules that have been developed and examines how each of these rules might flow from the various ethical theories.

CASUISTRY

Casuistry refers to a case-based system of ethical analysis (Jonsen, 1995). Artnak (1995) has summarized the techniques utilized in casuistical analysis as consisting of three features: typification, relationships to maxims, and certitude (Artnak, 1995). Typification refers to a comparison of the case at hand with the caregiver's past experiences, and identification of the similarities and differences between the instant case and those that preceded it. Relationships to maxims refers to reliance on "rules of thumb," that consider the characteristics of the situation at hand. Certitude refers to the certainty of the outcome in relationship to all of the alternative courses of action that are available. In essence, casuistry represents a "bottom-up" approach to the development of knowledge, rather than a "top-down" approach, as is perceived to be the case with principlism (Arras, 1991).

Unlike principlism, which sets forth principles to be applied in specific cases, casuistry discovers ethical principles in the cases to be analyzed (Jonsen, 1986). Jonsen (1995: 241) has defined what makes a "case":

> A case is a confluence of persons and actions in a time and a place, all of which can be given names and dates. A case, we say, is concrete as distinguished from abstract because it represents the congealing, the coalescence, or the growingtogether (in Latin, *concrescere*) of many circumstances. Each case is unique in its circumstances, yet each case is similar in type to other cases and can, therefore, be compared and contrasted. Cases can be posed at various levels of concreteness. Some will be composed of quite specific persons, times, and places; others will describe an event or practice in more diffuse terms, such as the "case of the Bosnian war" or the "case of medical experimentation." I refer to cases of the latter sort as "great cases" ....

The principles thus derived are subject to revision because they are intimately linked to their factual surroundings, which vary between cases and over time. Casuistical analysis, then, "might be summarized as a form of reasoning by means of examples that always point beyond themselves" (Arras, 1991: 35). This is similar to the development of common law, which derives from the analysis of judicial opinions. Each subsequent case is examined based on similarities and differences with similar cases that preceded it. "The ultimate view of the case and its appropriate resolution comes, not from a single principle, nor from a dominant theory, but from the converging impression made by all of the relevant facts and arguments that appear....(Jonsen, 1995: 245). This can be contrasted with the principlistic approach, which is more analogous to a code-based system of law, in which a situation may be resolved by reference to a prior codification of rules pertaining to such situations. Arras (1991) has argued that the emphasis on the development of principles from cases, rather than the reverse, requires the utilization of real cases, rather than hypothetical ones. Hypothetical ones, he argues, tend to be more theory-driven, rather than practice-driven.

There is a specific method to effectuate case comparison. First, the mid-level principles relevant to a case and role-specific responsibilities are identified. Potential courses of action must be determined. Third, the case is compared to others that are similar. Fourth, for each identified course of action, a case must be identified in which the option under consideration is justified. These justified cases are termed "paradigms" (Kuczewski, 1994). A paradigm case is

> [a] case in which the circumstances were clear, the relevant maxim unambiguous and the rebuttals weak, in the minds of almost any observer. The claim that this action is wrong (or right) is widely persuasive. There is little need to present arguments for the rightness (or wrongness) of the case and it is very hard to argue against its rightness (or wrongness). (Jonsen, 1991: 301)

New cases are then compared to a set of paradigm cases and a paradigm that is similar to the new case is identified. This identified similar paradigm then serves as a guide for action in the new case (Kuczewski, 1994; see DuBose and Hamel, 1995). A paradigm is "the ultimate justification of moral action....(Kuczewski, 1994: 105), although it must be recognized that some paradigms may be more stable than others.

Jonsen and Toulmin (1988) have characterized the work of the National Commission for the Protection of Human Subjects of Biomedical and Behavioral Research as casuistic in nature. The commissioners hail from a variety of academic and nonacademic disciplines. They analyzed paradigmatic cases involving harm and fairness and then extended their analysis to more complex situations raised by biomedical research, essentially utilizing an incremental approach to the examination of difficulties encountered in research.

Arras (1991) has identified numerous strengths of casuistry. First, this method encourages the development of detailed case studies, because their facts are critical not only in resolving the issue(s) at hand, but also in the resolution of case situations that do not yet exist. Second, the case-based approach encourages reliance not merely on a single case, but rather on a sequence of cases that relate to a central theme. Third, this strategy will require in-depth examination of exactly what issues are being raised in a particular situation.

These same characteristics, however, are weaknesses as well as strengths. For instance, it is unclear what situations and how a situation is to constitute a "case" meriting in-depth examination. Second, casuistry does not provide guidance on how detailed a case must be to be adequate, since a case necessarily involves multiple perspectives, numerous facts, and various issues. Third, the resolution of new situations requires reference to previous situations that embody similar issues and/or facts, but no guidance is available as to how to group cases thematically (Arras, 1991).

Other criticisms of casuistry have been voiced. Some critics have maintained that the casuistic ideal of reaching consensus on issues raised is illusory (Emanuel, 1991). Others assert that because casuistry does not depend on "metaphysical realism," "it substitutes current opinions, commonly held values, and prejudices..." and consequently reinforces the status quo (Kuczewski, 1994: 106). Jonsen, however, argued in response that the weight of a particular value should not depend on the shared understanding of the community, but instead should depend on the circumstances of the particular case.

COMMUNITARIANISM

Communitarianism is premised on several themes: the need for a shared philosophical understanding with respect to communal goals and the communal good, the need to integrate what is now fragmented ethical thought, and the need to develop "intersubjective bonds that are mutually constitutive of [individuals'] identities" (Kuczewski, 1997:3). Sandel (1982: 172) has explained the communitarian perspective:

> In so far as our constitutive self-understandings comprehend a wider subject than the individual alone, whether a family or tribe or city or class or nation of people, to this extent they define a community in the constitutive sense. And what marks such a community is not merely a spirit of benevolence, or the presence of communitarian values, or even certain "shared final ends" alone, but a common vocabulary of discourse and a background of implicit practices and understandings.

Etzioni (1998) has distinguished between the "old communitarians" and the "new communitarians." The former are characterized by their emphasis on the significance of social forces and bonds (Etzioni, 1998); the latter focus on the balance between social forces and the individual, between community and autonomy, between the common good and liberty, and between individual rights and social responsibility (The Responsive Communitarian Platform: Rights and Responsibilities, 1998: xxv). The new communitarians, then, are concerned with a dual danger: the society whose communal foundations are deteriorating and the society in which individual freedoms are negated. New communitarians advocate the promotion of pro-social behavior through persuasion, rather than through coercion (Etzioni, 1998).

Unlike principlism, which focuses on the rights of the individual, communitarianism examines communal values and relationships and attempts to ascertain which are present and which are absent. Where communitarianism emphasizes the need for a common vocabulary and shared understanding, casuistry rejects such a foundation, arguing that it is the "breakdown of tradition [that] forces reexamination of particular instances of action and a return to concrete practical reasoning...." (Kuczewski, 1982: 61; see Jonsen, 1980: 163-164).

One difficulty of communitarianism lies in the ability to define what constitutes the relevant community. Etzioni (1998: xiii-xiv) has offered the following:

> Communities need not be geographically concentrated....
> Communities are not automatically or necessarily places of virtue. Many traditional communities that were homogenous, if not monolithic, were authoritarian and oppressive. And a community may lock into a set of values that one may find abhorrent....
>
> However, contemporary communities tend to be new communities that are part of a pluralistic web of communities. People are, at one and the same time, members of several communities, such as professional, residential and others. They can, and do, use these multi-memberships . . . to protect themselves from excessive pressure by any one community.
>
> What is the scope of communities? It is best to think about communities as nested, each within a more encompassing one...Ultimately, some aspire to a world community that would encapsulate all people. Other communitarians object to such globalism and suggest that strong bonds and the moral voice, the essence of communities, mainly are found in relatively small communities in which people know one another, at least to some extent, as in many stable neighborhoods.

FEMINIST ETHICS

Feminine or Feminist?

A distinction has been made between a feminine and a feminist approach to ethics:

> "Feminine" at present refers to the search for women's unique voice and most often, the advocacy of an ethics of care that includes nurturance, care, compassion, and networks of communications. "Feminist" refers to those theorists, whether liberal or radical or other orientation, who argue against patriarchal dominations, for equal rights, a just and fair distribution of scarce resources, etc. (Sichel, 1991: 90)

Similarly, Sherwin (1992b: 42-43) has distinguished between feminine ethics and feminist ethics. The former "consists of observations of how the traditional approaches of ethics fail to fit the moral experiences and intuitions of women," while the latter refers to "[the application of] a specifically political perspective and . . . suggestions for how ethics must be revised if it is to get at the patterns of dominance and oppression as they affect women." Walker (1998: 22) explains what feminist theory does that others do not do:

> What feminists show is not that moral philosophy is simply mistaken in its claims to represent moral life. Rather, feminist critiques show how moral philosophers have in fact represented, in abstract and idealized theoretical forms, aspects of the *actual* positions and relations of *some* people in a certain kind of social order. This social order is the kind where the availability of these positions depends on gender, age, economic status, race, and other factors that distribute powers and forms of recognition differentially and hierarchically. Dominant moral theories depict the self-images, prerogatives of choice, required patterns of moral reasoning and anticipated forms of accountability of *some* people in societies like ours; those placed in certain ways, not just in any or every way.

Related to the distinction between feminine and feminist ethics is the issue of a feminine versus feminist consciousness and the etiological/developmental theories offered in support of each. Certain personality traits, such as compassion, caring, and nurturing, have traditionally been associated with women. Three theories have been offered to explain this association. Whitbeck (1984) has argued that the biological experiences of women, such as pregnancy and nursing, shape the woman's consciousness and her perception of herself as a caregiver, at first to the child and later, in a more extended manner, to the larger community. Chodorow's (1978) explanation is routed in the differences in the psychosexual development of girls and boys, whereby girls are able to form an identity while remaining emotionally linked to their mothers, while boys must disassociate from their mothers in order to form a masculine self-identity. Consequently, she argues, women are able to develop and value an ability to form relational networks and relationships, whereas men are more likely to value their ability to function alone. A third view attributes feminine consciousness to women's exclusion from the public world, where men speak to each other about professional and public affairs, and their relegation to the private space of the home, where they listen to the voices of others—babies, teens, and the elderly (Tong, 1993).

Tong (1993) has pointed out the difficulties with each of these approaches. The first explanation essentially posits that women are biologically programmed to give care and men are similarly programmed to receive it. Change, therefore, cannot be expected. The psychosexual approach forces one to question the wisdom of promoting women's dependence and men's independence. The third hypothesis potentially allows greater evolution in gender roles through the increased movement of women from the private to the public sphere. However, in so doing, it is unclear whether the positive aspects of women's feminine consciousness will or can be retained.

In contrast, feminist consciousness encompasses not only the fact of women's subordination in various forms, but also the elimination of this status. This consciousness has assumed various forms.

Liberal Feminism

Liberal feminists assert that women are restrained from entrance and success in the "public world" as a result of various legal and customary obstacles. Men and women can be equals only after women are afforded the same educational and occupational opportunities as men (Friedan, 1974).

Marxist Feminism

Marxist feminists claim that capitalism, not merely the larger society as claimed by the liberal feminists, is responsible for the oppression of women. The replacement of the capitalist structure with one that is socialist will allow women and men to be economic equals, facilitating the development of political equality (Barrett, 1980).

Radical Feminism

Radical feminists attribute women's subordination to their reproductive roles and responsibilities and to "the institutionalization of compulsory heterosexuality," rather than to only economic, educational, or occupational inequality (Rich, 1980: 648-49). Consequently, change requires the elimination of the social, cultural, legal, political, and economic institutions and infrastructures that have contributed to women's subordination. Reproductive technologies, such as artificial insemination and in vitro fertilization, are viewed with skepticism, as further encouraging women's complete self-sacrifice to the benefit of others.

Psychoanalytic Feminism

Psychoanalytic feminism focuses on sexuality and roles, maintaining that systems of dual parenting and careers are needed so that children are not routinized to images of a working father and a nurturing mother (Mitchell, 1974).

Socialist Feminism

Socialist feminism has attempted to identify common themes across these various perspectives. Mitchell (1971) has delineated four functions that disproportionately impact on a woman's condition: production, reproduction, sexuality, and the socialization of children. The function and status of women in each of these realms must change as a precondition to women's equality with men.

Although distinct, each of these theories focuses on "a methodology of feminist thought" (Tong, 1997: 93). Sherwin (1992b: 32) has observed:

> I believe we must expect and welcome a certain degree of ambivalence and disagreement within feminist theorizing. Contemporary feminism cannot be reduced to a single, comprehensive, totalizing theory.

Foundations of Feminine and Feminist Ethics

Rousseau had asserted that men and women are, by nature, different and, consequently, had different educational needs. Males were to be educated to be rational, whereas women were to be trained to be patient and flexible. Although women could be trained to develop masculine traits, he argued that they should not be. Women were incapable of autonomy. Wollenstone (1988) disputed the premises underlying Rousseau's conclusions about men and women, and asserted that women could develop the qualities that would allow them to be independent.

Mill (1911: 32) recognized that the virtues that were perceived to differ between men and women were the result of society's dictates:

> All women are brought up from the very earliest years in the belief that their ideal of character is the very opposite to that of men; not self-will and government by self-control, but submission and yielding to the control of others. All the moralities tell them that it is the duty of women, and all the current sentimentalities that it is their nature, to live for others, to make complete abnegation of themselves, and to have no life but in their affections.

Women, Mill argued, could not be true moral agents without education and suffrage.

Stanton (quoted in Buhle and Buhle, 1978: 325-326) asserted that men and women were created equal. She maintained, however, that women's rights were subordinated to their duties, unlike men, whose rights were not subordinated to their duties. Stanton recognized that women's continuing self-sacrifice would ultimately negatively impact on women's ability to gain political and economic power (Stanton, 1993: 131).

Feminine Approaches to Ethics

The Ethic of Care

The ethic of care derives from empirical observations which found that men tend to resolve situations utilizing an ethic of rights, with an emphasis on fairness, while women tend to rely on an ethic of caring that focuses on needs, care, and the prevention of harm (Gilligan, 1982). Gilligan, then, did not dispute that there exist differences between men and women. Rather, Gilligan maintained that society had placed a greater value on individual achievement, thereby devaluing the caretaking roles fulfilled by women:

> The very traits that have traditionally defined the "goodness" of women, their care for and sensitivity to the needs of others, are those that mark them as deficient in moral development. The infusion of feeling into their judgments keeps them from developing a more independent and abstract ethical conception in which concern for others derives from principles of justice rather than from compassion and care (Gilligan, 1977: 484).

She argued that two different modes of social experience must be recognized:

> The failure to see the different reality of women's lives and to hear the differences in their voices stems in part from the assumption that there is a single mode of social experience and interpretation. By positing instead two different modes, we arrive instead at a more complex rendition of human experience. (Gilligan, 1982: 173-174)

The ethic of care, then, rejects the cognitive emphasis of other approaches to ethical analysis and emphasizes the moral role of the emotions. The detachment inherent in the cognitive approaches is criticized precisely because it fails to recognize the attachment inherent in relationships.

Gilligan (1982) delineated three levels in the development of an ethic of care and responsibility: orientation to individual survival, goodness as equated with self-sacrifice, and nonviolence. There are two transitions between levels. The first reflects movement from selfishness to responsibility for others and occurs between the first two levels. The second transition, occurring between the second and third levels, reflects increasing attention to oneself as well as to others.

Gilligan's paradigm is to be contrasted with that of Kohlberg's model of the development of moral judgment, which Gilligan argued had represented men's development, not human development. In contrast to Gilligan's phenomenological orientation, Kohlberg's model emphasized rationality. Gilligan's paradigm emphasized the self as connected and attached, relational, whereas Kohlberg viewed the self as individual and separate. Kohlberg's (1976; Colby and Kohlberg, 1987) model reflected three levels and six stages. The levels, from lowest to highest, were termed the preconventional, the conventional, and the postconventional or principled. Stages 1 and 2, that of heteronomous morality and individualism, instrumental purpose and exchange, consist of compliance with rules in order to avoid punishment and compliance with rules when it is for one's own benefit. The third and fourth stages, placed at level 2, consist of mutual interpersonal expectations, interpersonal conformity, and the maintenance of social system and conscience. Adherence to rules during these stages is a function of meeting the expectations of others or of a pre-existing agreement. Stage 5 focuses on individual rights and social contract, while stage 6 entails universal ethical principles.

Adherence to and compliance to rules rests on the existence of a social contract and self-chosen ethical principles.

Blum (1993) has cogently enumerated seven major differences between the Kohlberg and Gilligan formulations of morality.

1. Gilligan's moral self is defined by its historical connections and relationships, whereas Kohlberg's moral self attempts to achieve an impersonal standpoint that defines the moral point of view.
2. Gilligan conceives of the self, the other, and the situation as particularized, whereas Kohlberg does not.
3. Gilligan views the acquisition of knowledge of the other person as a complex task that requires reliance on specifically moral capacities. Kohlberg emphasizes, in contrast, the acquisition of knowledge of the other as an empirical process.
4. Gilligan conceives of the self in relation to others; those relationships are not necessarily as result of choice. Kohlberg, though, views the individual as an autonomous, independent agent.
5. Gilligan's concept of morality involves a consideration of cognition, emotions and action, whereas Kohlberg views morality as a function of rationality, with emotions playing only a secondary, if any, role.
6. Gilligan views the appropriateness of action in a particular situation as nonsubjective and premised on standards of care and responsibility. Kohlberg, however, views the principles of right action as universal.
7. For Gilligan, moral action expresses and sustains the connections with others, while Kohlberg's ultimate moral concern is with morally right action and principle.

Tong (1996) has argued that a care-oriented ethic is not in and of itself neglectful of issues relating to gender inequity. Rather, an ethic of relationships and nurturance should encourage all human beings to care for each other and facilitate women's liberation from oppressive systems and structures (Manning, 1992). Such feminists have been termed "cultural feminists" (West, R.C., 1988).

The ethic of care has been criticized on a number of bases. First, Nicholson (1993) has argued that our understanding of reality need not be limited to two different modes. Tronto (1993) has asserted that the advocates of an ethic of care have not adequately explored and explained the assumptions on which this moral position is premised.

Relational Ethics

Noddings (1984), like Gilligan, has criticized traditional ethical theories for their undervaluing of caring and their counterintuitive approach to issues arising in the context of relationships. Noddings goes further than does Gilligan, however, and argues that an ethics of care is better than, not only different from, an ethic of justice. Noddings' objections to universalism are moderated by her observation that the caring attitude that underlies her ethical view is, in itself, universal.

Noddings (1984: 5) maintains there exists a natural caring:

> The relation of natural caring will be identified as the human condition that we, consciously or unconsciously, perceive as "good." It is that condition toward which we long and strive, and it is our longing for caring—to be in that special relation—that provides the motivation for us to be moral. We want to be moral in order to remain in the caring relation and to enhance the ideal of ourselves as one-caring.

She argues that an ethic of caring can be taught as easily as an ethic of rules and principles. Moral development does not require the replacement of natural caring with ethical caring.

Noddings asserts that relational ethics encompasses two types of virtues. The first is that set of virtues that belong to the relationship of the people involved. The second set consists of those virtues that belong to the individuals involved in the relationship, such as honesty. Noddings frames most situations in terms of relational dramas, arguing, for instance, that even a decision relating to euthanasia should be made in consultation with those who will be affected by the patient's suffering and dying. There is an underlying assumption that such consultations and discussions will result in a decision by consensus.

Tong (1993) and others (Card, 1990; Hoagland, 1991) have criticized Noddings relational ethics on a number of grounds. Tong disputes the notion that consensus will be reached through discussions involving members of a relational network. Tong notes that most individuals are members of networks that are not altogether healthy ones. Hoagland has criticized Noddings' (1) reliance on unequal relationships as the basis for her assertions; (2) assumption that the individual providing the care is in the best position to know what is good for the person receiving the care; (3) assumption that inequalities in abilities, rather than power, necessarily render a relationship unequal; and (4) advisory that the cared-for blindly trust their caregivers, pointing out that this would result in the vulnerability of those receiving the care and potentially subject them to abuse. Hoagland further advises that Noddings has failed to distinguish between receptivity to caring and reciprocity of caring and, by emphasizing receptivity, has failed to incorporate into her model the kind of respect that is truly necessary for a moral relationship.

Card (1990) has observed that the virtue of justice is an intrinsically valuable moral virtue, as is caring, and serves as a basis for the clarification and delineation of our obligations to others. Both Card and Hoagland have asserted that Nodding emphasizes self-sacrificial love (*agape*) to the detriment of the caregiver:

> In direct contrast to eros, which is self-centered, agape is other-centered. The caring of agape always moves away from itself and extends itself unconditionally. Certainly Nel Nodding's analysis is that caring moves away from itself. However, I would add that since there are no expectations of the cared-for beyond being acknowledged by the one-caring, since my ethical self can emerge only through caring for others, since withdrawal constitutes a diminished ideal, and since there is allegedly no evaluation in receiving the other, one-caring extends itself virtually unconditionally (Hoagland, 1991: 257).

Feminist Approaches to Ethics

Purdy (1992) has delineated four tasks that can be accomplished by feminist ethics: (1) the provision of an emphasis on the importance of women and their interests; (2) the provision of a focus on issues especially affecting women; (3) the re-examination of fundamental assumptions; and (4) the incorporation of feminist insights from other fields into the field of ethics. Sherwin (1992a) assigns to feminist medical ethics the responsibility of developing conceptual models that will restructure the power associated with healing, to allow individuals to have the maximum degree of control possible over their own health.

The Content of the Discussion

Feminist ethics is often concerned with the content of the discussion (Warren, 1992), such as reproductive technologies and the rationing of medical care. Sherwin (1992a) has argued that feminist ethics frequently approaches situations from a general point of view, rather than examining specific applications only. For instance, while many ethicists approach the issue of abortion by weighing the relative importance of preserving life or protecting autonomy, feminist ethicists approach the issue of abortion by examining the difference that it will make in women's lives if they are free to decide to continue or not to continue each pregnancy. Sherwin (1992a: 25) concludes that "[i]ncreased reproductive technology generally means increased medical control."

Lesbian ethics has provided another view of what should be discussed and how. Hoagland (1988: 12) has described the focus of lesbian ethics as

> enabling and developing individual integrity and agency in relation to others...a self who is both separate and related, a self which is neither autonomous nor dissolved: a self in community who is one among many ...

Frye (1991) has interpreted Hoagland's call for lesbian ethics as a challenge to shift attention from what is good or right to deciding what should be the focus of attention.

The Participation of Women in Research

Feminist ethicists have long argued that the medical research agenda in the United States is determined with reference to those who are white, upper and middle class, and male (Rosser, 1992). The consequences of the resulting narrow perspective are troublesome: (1) hypotheses are developed and research conducted without reference to sex or gender, although the frequency of various diseases differs by sex; (2) some diseases which affect both sexes, such as coronary heart disease, are defined as male diseases, resulting in little research being conducted on women with those diseases; (3) research affecting primarily women has received a low funding priority; and (4) suggestions for research based on personal experiences of women have been ignored.

The exclusion of women from research results from a number of mechanisms, including eligibility criteria that specifically exclude women from participation in specific studies (Bennett, 1993) and reliance on gender-neutral eligibility that serve to exclude because they fail to take into account women's responsibility for the home and family (Merton, 1996). The exclusion of women has been defended with reference to the need for homogeneity among research participants to facilitate the research and statistical analysis, the potential liability that could result should a woman and/or her offspring be injured during the course of the research, and a belief that it is morally wrong to include women in studies because they may be, or may become, pregnant (Merton, 1992; Merton, 1996).
The one notable exception has been the inclusion of women in contraceptive research, which is

> directly related to men's interest in controlling production of children. Contraceptive research may permit men to have sexual pleasure without the production of children; research on infertility, pregnancy and childbirth has allowed men to assert more control over the production of perfect children and over an aspect of women's lives over which they previously held less power (Rosser, 1989: 128).

Each of the arguments against the inclusion of women is subject to refutation. First, research has indicated that drug metabolism and dose-response differ between men and women (Hamilton & Parry, 1983; Raskin, 1974). Consequently, it cannot be assumed that the results of clinical trials involving men are generalizable to women (Council on Ethical and Judicial Affairs, 1991). Second, the concept of homogeneity is a relative one that depends on our state of knowledge with respect to a specific health problem or population at a specific point in time (Levine, 1978). For instance, research involving diabetes must distinguish between types of diabetes. However, the usurpation of a woman's right to decide whether or not to participate in research in favor of the "rights" of her potential child is violative of respect for autonomy within the principlistic framework and perpetuates the oppression and subordination of women within the feminist paradigm. Roberts (1991: 1472) has argued that

> The right to bear children goes to the heart of what it is to be human. The value we place on individuals determines whether we see them as entitled to perpetuate themselves in their children. Denying someone the right to bear children—or punishing her for exercising that right—deprives her of a basic part of her humanity.

Genetic research has been of particular concern for some feminist ethicists. Because women are already deemed to be responsible for reproduction and family life in general, it is feared that women's reproductive options will become fewer as physicians insist, based on newly acquired knowledge, that genetic testing constitutes the standard of care (Asch and Geller, 1996). Some theorists visualize a devaluation of motherhood, as women are denied the role of *the* mother of their children as a result of new technologies. Corea (1985: 39) analogized the participation of women in such ventures to the trade of a prostitute: "While sexual prostitutes sell vagina, rectum, and mouth, reproductive-prostitutes will sell other body parts: wombs, ovaries, egg."

Because of these concerns, at least one ethicist has asserted that any ethical guidelines for preembryo research must involve women in their formation and in the formation of national policies relating to preembryo research and that the impact of proposed national policies on women as a group should be considered in their assessment (Strong, 1997). Additional issues that must be addressed include the replacement of preembryos used in research into a woman's uterus, the permissibility of creating preembryos for research purposes only, and the developmental stage to which preembryos or embryos may be maintained in the laboratory (Carson, 1997).

Researcher-Participant Communication

Clinical ethics has often focused on the doctor-patient relationship, which has been characterized as one that involves an imbalance of power and vulnerability (Peppin, 1994). The literature relating to the mode of communication between the physician and the patient in the clinical context is relevant, as well, to the communication between the researcher and the research participant

Using a principlistic approach to physician-patient communication, physician-patient communication presumably would encourage respect of the person, the provision of sufficient information to permit the patient to make an informed decision, support for the patient during the decisionmaking process, and mutual understanding. However, much of the communication between physicians and patients has been characterized as essentially a transfer of information ("information-transfer model")(Smith, 1996: 187). This involves the transfer of information from the patient to the physician to permit appropriate diagnosis and a subsequent transfer, in the form of explanations, from the physician back to the receiver-patient. The exchange focuses on the transfer of factual information. Such an emphasis, in conjunction with the authority granted to physicians, may lead a patient to believe that what the physician intended to be a factual assertion was actually a directive. This pattern has been characterized as an "interview," rather than a "conversation" (West, C., 1983: 76).

A communicative action model has been proposed as more appropriate and more adequate ethically than the information-transfer model. Unlike the information-transfer model, the communicative action model emphasizes action based on equality, mutuality, and consensus, which can only occur in the presence of real understanding. True consensus does not permit agreement based on persuasion (Fisher, 1986) or the selective provision of information due to physician-patient differentials in power and status stemming from differences in race and sex (Todd, 1983). As such, this model is better aligned with the principle of respect for autonomy and the legal and ethical requirement of informed consent. The communicative action model permits the patient to challenge statements, ask questions, and offer her own opinion. It also encourages the physician to view the patient as a "concrete other," rather than a "generalized other." When viewed as a concrete other, the patient is seen as a distinct individual with specific needs and wants. In contrast, the generalized other is seen as a "rational risk-taker, a bearer of rights, and an autonomous decision maker, giving or withholding informed consent to medical procedures" (Smith, 1996: 202).

The differences between these models is illustrated by the following clinical hypothetical. A young woman, recently married and two months pregnant, is told by her gynecologist that she has tested positive for the AIDS virus. The physician explains to her that there is a chance that her child will be born with HIV and that she may want to consider the possibility of an abortion. He also informs her, without inquiring into her particular circumstances, of the various mechanisms for HIV transmission. She interprets this sequence of information as the physician's hint to her that her husband must be having an affair. Additionally, she understands the doctor's suggestion of an abortion to be a "prescription" for what should be done, particularly because he followed it with information relating to HIV transmission.

The physician has succeeded in transferring necessary information to his patient, but does not realize that his patient's understanding of that information is not congruent with his intended message. If a communicative action model had been utilized instead of the information-transfer model, the patient would have been encouraged to ask questions and the physician would have engaged the patient in discussion in order to assess and enhance her level of understanding prior to her decision making. This communicative action model is most consistent with a feminist approach to bioethics in that it is concerned with the woman's needs and the potential prevention of harm.

A Critique

Feminist theory has been criticized as being underdeveloped, too contextual and hostile to principles, and overly confined to the private sphere of relationships (Beauchamp and Childress, 1994). The ethic of care has also been recognized as providing an important corrective to rights-based theory and a focus on impartiality, to the neglect of sensitivity and practical judgment (Beauchamp and Childress, 1994).

PRINCIPLISM

Central to principle-based theory is the existence of governing principles that enunciate obligations (Beauchamp & Childress, 1994.) The term principlism is often used to refer to four standard principles said to be derived from the Nuremberg Code, and further elucidated by the Helsinki Declarations. Principlism is, in essence, the overriding approach utilized in the United States. These principles are respect for autonomy, nonmaleficence, beneficence, and justice (Beauchamp & Childress, 1994).

Respect for autonomy encompasses the concept of informed consent. In turn, informed consent requires competence or capacity, voluntariness, disclosure of information, and understanding. Competence, or the lack of it, is often determined by reference to one of three standards: (1) the ability to state a preference, without more; (2) the ability to understand information and one's own situation; and (3) the ability to utilize information to make a life decision (Appelbaum & Grisso, 1988; Appelbaum, Lidz & Meisel, 1987). Voluntariness refers to the individual's ability to consent or refuse a treatment or procedure or participation without coercion, duress, or manipulation (Beauchamp & Childress, 1994).

The concept of disclosure focuses on the provision of information to the patient or research participant. Ethically, the health care provider must disclose the facts that the patient or research participant would consider important in deciding whether to consent or to withhold consent, information that the health care provider believes is material, the recommendation of the health care provider, the purpose of the consent, and the scope of the consent, if given (Beauchamp & Childress, 1994). Legally, additional information may be required. For instance, legally a physician may be required to disclose personal interests that may affect his or her judgment, whether or not those interests are related to the patient's health (*Moore v. Regents of the University of California*, 1990).

Understanding is related to the disclosure of information in a way that can be understood. Studies have shown, for instance, that information provided to prospective participants in research studies is often written at a level above the participants' educational level (Hammerschmidt & Keane, 1992; Meade & Howser, 1992) and often includes unfamiliar words, long words, and long sentences (Rivera, Reed, & Menius, 1992). Understanding may also be impeded in situations where the patient or research participant comprehends the information but refuses to accept the information. For instance, a patient may refuse to consent to an HIV test where she intellectually understands what behaviors may subject an individual to an increased risk of transmission, but believes that such a test is unnecessary for her because she is not ill and she believes that HIV-infected persons must look and feel sick.

Nonmaleficence refers to the obligation to refrain from harming others. Conversely, the principle of beneficence "refers to a moral obligation to act for the benefit of others""(Beauchamp & Childress, 1994: 260). This must be distinguished from benevolence, which refers to the character trait of one who acts for the benefit of others. Although some beneficent acts may be admirable, they are not necessarily obligatory, such as the donation of blood or of an organ to another. The principle of justice refers to the distribution of the benefits and burdens, e.g. of health care and of research. How those benefits and burdens should be distributed is the subject of intense and ongoing debate.

The principlistic approach has been criticized on a number of grounds. First, the principles themselves can be in conflict in specific situations and they provide little or no guidance in resolving such conflicts. For instance, the principle of autonomy would suggest that a woman has the right to decide whether or not to participate in a clinical trial for a new drug designed to reduce transmission of a sexually transmitted disease, based upon receipt of all information material to that decisonmaking process. Beneficence, however, would argue that she should not be permitted to participate because of the unknown and unknowable risks to any future unborn children. Second, principlism does not have a systematic theory as its foundation (Green, 1990). Clouser and Gert (1990) and Winkler (1993) have argued, for instance, that principlism borrows its four principles from others, without really integrating them into a unified theory: autonomy from Kant, justice from Rawls, beneficence from Mills, and nonmaleficence from Gert. Third, various critics have argued that principlism is too individualistic, rights-focused, and rationalistic and is exceedingly narrow in its understanding of various religious and cultural frameworks (Clouser & Gert, 1990; DuBose, Hamel, & O'Connell, 1994).

DEONTOLOGY

Deontology values actions based upon the underlying intent: are the actions motivated by the person's intent to perform his or her duty, because it is his or her duty? Actions, therefore, can be morally worthy, unworthy, or nonworthy. Unworthy actions are those that are performed without regard to relevant moral law. Those that are nonworthy are those that are performed in conformity with moral law, but for a motive other than adherence to moral law. For instance, a researcher may conform his or her actions as a researcher to moral law because it is a duty; this is an example of moral action. However, a researcher may conform to moral law, not because he or she is obliged to do so morally, but because he or she is afraid of the legal consequences if the precepts are violated. This is an example of morally nonworthy action. A researcher who violates the precepts relevant to conducting research is engaging in morally unworthy action.

Kant (trans. Paton, 1956: 74-75) has enumerated three conditions, known as the "Categorical Imperative," that make a rule a moral rule:

1. Act only on that maxim through which you can at the same time will that it should become universal law;
2. Act in such a way that you always treat humanity, whether in your own person or in the person of another, never simply as a means but always at the same time as an end; and
3. Never . . . perform an action except on a maxim such as can also be a universal law, and consequently such that the will can regard itself as at the same time making universal law by its maxim.

From an analysis of these conditions, Kant concluded that each person must regard him- or herself and other persons as having unconditional worth, as contrasted with the worth of objects, which is conditioned upon the value that we assign to that object. Individuals should not be treated merely as objects. For instance, the researcher should not treat the participants in his or her study as merely objects upon which he or she can conduct the experiment. Rather, the participants are to be treated with respect.

Deontological constraints are often framed negatively, *e.g.* do not lie. They also are framed and directed narrowly and are bounded:

> In every case the [deontological] norm has boundaries and what lies outside those boundaries is not forbidden at all. Thus lying is wrong, while withholding a truth which another needs may be perfectly permissible—but that is because withholding a truth is not lying. (Fried, 1978: 9-10)

Certain actions are deemed to be wrong because of the nature of the acts themselves:

> Common moral intuition recognizes several types of deontological reasons—limits on what one may do to people or how one may treat them. There are special obligations created by promises and agreements; the restrictions against lying and betrayal; the prohibitions against violating various individual rights, rights not to be killed, injured, imprisoned, threatened, tortured, coerced, robbed; the restrictions against imposing certain sacrifices on someone simply as a means to an end; and perhaps the special claim of immediacy, which makes distress at a distance so different from distress in the same room. There may also be a deontological requirement of fairness, of even-handedness or equality in one's treatment of people. (Nagel, 1986: 176)

However, in contrast with utilitarianism, discussed below, what is "right," in contrast to what is "wrong," is not always easily discernible within a deontological framework. Whereas utilitarianism accepts as "right" that which maximizes overall utility, deontology has as its precept that

> [o]ne cannot live one's life by the demands of the domain of the right. After having avoided wrong and doing one's duty, an infinity of choices is left to be made (Fried, 1978: 13).

Deontology has been criticized for it lack of guidance in the resolution and/or prioritization of absolute, but conflicting rules. Additionally, Kantian deontology stresses that feelings neither add nor detract from the moral worthiness of actions. Ross as attempted to address the situation involving conflicting rules by distinguishing between actual moral duty and *prima facie* ("at first sight") moral duty. Ethical conflicts arise due to the conflict between two *prima facie* duties. Moral intuitions, Ross asserts, not only assist us in deciding which of our prima facie duties are also our actual moral duties, but help us to recognize the *prima facie* duties (Tong, 1997).

UTILITARIANISM

The theory of utilitarianism is premised on the idea of utility: that the "aggregate welfare is the ultimate standard of right and wrong" (Reiman, 1988: 41). The "right" course of action is determined by summing the "good" consequences and the "bad" consequences to welfare that may result from each alternative course of action and selecting that course of action that appears to maximize the "good" consequences to welfare. Utilitarianism, then, values an action based upon its utility-maximizing consequences and has, as a result, been known as consequentialism (Williams and Smart, 1973).

How to measure gains and losses to welfare, however, is far from simple and, to a great degree, depends on which values are most important and how they are to be weighed. For instance, the maximization of good can be premised on the value of happiness, *i.e.*, whichever course of action produces the greatest degree of happiness, or it can refer to the maximization of goods valued by rational persons. Mill (1863, reprinted 1998: 124) explained what the maximization of happiness means:

> [T]he ultimate end, with reference to and for the sake of which all other things are desirable—whether we are considering our own good or that of other people—is an existence exempt as far as possible from pain, and as rich as possible in enjoyments, both in point of quantity and quality; the test of the quality and the rule for measuring it against quantity being the preference felt by those who, in their opportunities of experience, to which must be added their habits of self-consciousness and self-observation, are best furnished with the means of comparison.

Quality was to be assessed as follows:

> If I am asked what I mean by difference in quality of pleasures, or what makes one pleasure more valuable than another, merely as a pleasure, except its being in greater amount, there is but one possible answer. Of two pleasures, if there be one to which all or almost all who have experience of both give a decided preference, irrespective of any feeling of moral obligation to prefer it, that is the more desirable pleasure. If one of the two is, by those who are competently acquainted with both, placed so far above the other that they prefer it, even though knowing it to be attended with a greater amount of discontent, and would not resign it for any quantity of the other pleasure which their nature is capable of, we are justified in ascribing to the preferred enjoyment a superiority in quality so far outweighing quantity as to render it, in comparison, of small account. (Mill, 1863, reprinted 1998: 123)

Mill's explanation fails, however, to explain how much experience is required or sufficient to assess the quality of a particular end and how one person's competence is to be judged. Mill did, however, address the difficulty in assessing even what is to be accepted as "good":

> Questions of ultimate ends are not amenable to direct proof. Whatever can be proved to be good must be so by being shown to be a means to something admitted to be good without proof. The medical art is proved to be good by its conducing to health; but how is it possible to prove that health is good? (1863, reprinted 1998; 119)

Unlike Mill, Bentham (1962) proposed that individuals consider intensity, duration, certainty, propinquity, fecundity, purity, and extent, in assessing the utility of specified actions for specified individuals.

The role of rules in utilitarianism is somewhat controversial. Some utilitarians ("act utilitarians") would argue that rules provide a rough guide, but do not require adherence where the greatest good in a particular circumstance may result from breach of the rule. Others emphasize the importance of the rule in maximizing the "good" consequences, as demonstrated by one utilitarian in discussing the importance of truth-telling in the context of the physician-patient relationship:

> The good, which may be done by deception in a few cases, is almost as nothing, compared with the evil which it does in many, when the prospect of its doing good was just as promising as it was in those in which it succeeded. And when we add to this the evil which would result from a general adoption of a system of deception, the importance of a strict adherence to the truth in our intercourse with the sick, even on the ground of expediency, becomes incalculably great (Hooker, 1849:357).

Smart (1961) has advocated a third possibility, that of sometimes relying on rules.

Utilitarianism has been criticized on numerous grounds. Beauchamp and Childress (1994) have argued that utilitarianism appears to permit blatantly immoral acts where such acts would maximize utility. As a result, the appropriateness of including utilitarians on institutional review boards, discussed below, has been called into question (Reiman, 1988). Tong (1993) has detailed the difficulty inherent in the utilitarian perspective. Utilitarians do not, for instance, want to defend preferences that are discriminatory. Consequently, they may attempt to distinguish between acceptable and unacceptable preferences, or rational and irrational preferences, classifying those that are discriminatory as unacceptable and/or irrational. However, distinguishing between rational/irrational and acceptable/unacceptable may be equally difficult. Additionally, a utilitarian perspective requires that an individual subvert his or her preference to the larger good, something that many may be unwilling to do. Williams and Smart (1973) have noted the difficulty in even establishing causality to determine utility, *e.g.*, whether a particular action is related to a particular consequence, or whether a situation is so attenuated from an action that it cannot be said to be a consequence of it.

Donagan (1968) has asserted that utilitarianism fails to distinguish between those actions that are morally obligatory and those that are performed based on personal ideals and are above and beyond the call of moral obligation. In a similar vein, Williams and Smart (1973: 97) have noted that utilitarianism essentially creates negative responsibility:

> It is because consequentialism attached value ultimately to states of affairs, and its concern is with what states of affairs the world contains, that it essentially involves the notion of *negative responsibility*: that if I am ever responsible for anything, then I must be just as much responsible for things that I allow or fail to prevent, as I am for things that I myself, in the everyday restricted sense, bring about. Those things must also enter my deliberations,

as a responsible moral agent, on the same footing. What matters is what states of affairs the world contains, and so what matters with respect to a given action is what comes about if it is done, and what comes about if it is not done, and those are questions not intrinsically affected by the nature of the causal linkage, in particular by whether the outcome is partly produced by other agents.

Despite its obvious shortcomings, the utilitarian perspective has various strengths. First, the emphasis placed on consideration of the consequences potentially serves to maximize beneficence, when it is the good of all that is the ultimate goal. As Sen (1987: 75) has observed, "Consequentialist reasoning may be fruitfully used even when consequentialism as such is not accepted. To ignore consequences is to leave an ethical story half told." Second, because utilitarianism appears to demand from individuals extraordinary service, the perspective challenges individuals to rise above an ordinary level of function.

CONTRACT-BASED ETHICS

Rawls has suggested that the ideal of a social contract may be a useful basis for discussion, although the reality of one is unlikely. Governments should be judge, he suggests, by a comparing their merits to the terms of a social contract which would be developed by rational individuals attempting to establish a new government. Rawls proposed that, to begin this process, the worst possible situation should be envisioned. Rawls (1971: 12) has described how the principles of justice would evolve:

> Among the essential features of this situation is that no one knows his place in society, his class position or social status, nor does anyone know his fortune in the distribution of natural assets and abilities, his intelligence, strength, and the like.... [They] do not know their conceptions of the good or their special psychological propensities. The principles of justice are chosen behind a veil of ignorance.

Justice, then, will be as follows.

> First, each person is to have an equal right to the most extensive liberty compatible with a similar liberty for others. Second, social and economic inequalities are to be arranged so that they are both (a) reasonably expected to be to everyone's advantage and (b) attached to positions and offices available to all. (Rawls, 1971: 83)

Rawls' formulation has, however, been criticized for difficulties associated with its application. His concept of distributive justice, through which society's benefits and burdens are distributed, cannot be implemented until the first precept underlying justice is effectuated.

VIRTUE ETHICS

Unlike deontology, which focuses on obligation; principlism, which focuses on the application of principles; and utilitarianism, which addresses consequences, virtue ethics focuses on the agents who make choices and perform actions. Loewy (1996: 29) succinctly summarized the precepts underlying virtue ethics:

> "Virtue" is used here in the sense of competence in the pursuit of moral excellence....To Plato, virtue was synonymous with excellence in living a good life, and such excellence could be attained by practice. Vice, Plato believed, was not so much caused by moral turpitude as it was the result of ignorance: One either lacked knowledge or lacked the ability to reason properly. To Aristotle and, later, Aquinas, virtue was a disposition to act on the right way. Aristotle saw in practical terms virtue was the result of a balance among intellect, feeling, and action. "Virtue" was a state of character and the result of practice. In turn, practice resulted in habit so that the "virtuous" man could be counted on to act justly.

Aristotle's "practical wisdom" embodies the ability to identify which ends are worthy of pursuit of both specific and general goods, such as the success or development of an individual within a community. The absence of practical wisdom results in lack of moderation and a tendency to deficiency or excess (Aristotle, trans. Wardman, 1963).

Beauchamp and Childress (1994) have asserted that principles and virtues are compatible. For instance, the principles of respect for autonomy, nonmaleficence, beneficence, and justice correlate to the virtues of respectfulness, nonmalevolence, benevolence, and justice or fairness, respectively. Rules relating to veracity, confidentiality, privacy, and fidelity correspond to the virtues of truthfulness, "confidentialness," respect for privacy, and faithfulness. Ideals of action, such as exceptional forgiveness, are similar to ideals of virtue. Beauchamp and Childress have criticized virtue theory, however, arguing that character judgments are often less essential than rules or principles in guiding interactions and that virtue is inadequate to justify or explain actions. Other critics have noted that Aristotle's concept of practical wisdom relies heavily on a consensus regarding what is truly valuable, which may be difficult to achieve.

PRAGMATISM

Pragmatists view rules and principles in relation to their actual effectiveness. Dewey, one of the foremost pragmatists, conceived of ethics as process of inquiry, which requires decisionmaking in the face of imperfect knowledge of the facts, circumstances, and potential consequences. Dewey observed that our ethics changes over time, in response to changes in conditions around us. Consequently, conflicts among ethical theories reflect varying social needs over time. Moral inquiry requires the assessment of a situation from varying disciplines and diverse perspectives and the selection of values that will assist in the interpretation of the facts and facilitate the development of individuals' and communities' integrity.

CHAPTER SUMMARY

This chapter reviewed various approaches that can be utilized to resolve ethical dilemmas. Some methods of analysis, such as deontology and principlism, emphasize obligation and justice while others, such as feminist ethics, emphasize the interrelatedness of individuals. Some theories can be classified as "top down," resolving situations through the application of governing principles, while others, such as casuistry, can be considered "bottom up," deriving principles from an examination of the situation at hand. Each approach has its strengths and weaknesses, and it is possible that the resolution of different situations may call for different approaches. Chapter 3 discusses the derivation of principles and rules from these theories and their application, through various international and national guidelines and regulations.

EXERCISE

You are interested in using discarded fetal tissue in your research on various types of genetic diseases. Assume for the purpose of this question that there are no legal restrictions on the use in research of fetal tissue from abortion. How might this type of research be viewed by proponents of each of the ethical theories discussed above? In your response, consider how each of the following factors might or might not impact on your conclusion: (1) the voluntariness of the pregnancy; (2) the voluntariness of the abortion; (3) the underlying reason for the abortion; (4) the stage of pregnancy during which the abortion occurs; (5) the disposal of the tissue as human waste if no research is conducted; and (6) the status of women.

References

Appelbaum, P.S. & Grisso, T. (1988). Assessing patients' capacities to consent to treatment. *New England Journal of Medicine, 319*, 1635-1638.

Appelbaum, P.S., Lidz, C.W., & Meisel, A. (1987*). Informed consent: Legal theory and clinical practice.* New York: Oxford University Press, 1987.

Aristotle, Nicomachean Ethics, book 2. A.E. Wardman (Trans.). In R. Bambrough (Ed.). (1963). *The Philosophy of Aristotle.* New York: Mentor Books.

Arras, J.D. (1991). Getting down to cases: The revival of casuistry in bioethics. *The Journal of Medicine and Philosophy, 16*, 29-51.

Artnak, K.E. (195). A comparison of principle-based and case-based approaches to ethical analysis. *HEC Forum, 7*, 339-352.

Asch, A. & Geller, G. (1996). Feminism, bioethics, and gender. In S.M. Wolf (Ed.), *Feminism and bioethics: Beyond Reproduction* (pp. 318-350). New York: Oxford University Press.

Barrett, M. (1980). *Women's Oppression Today.* London, England: Verso.

Bentham, J. (1962). An introduction to the principles of morals and legislation. In *Utilitarianism and Other Writings.* New York: The New American Library.

Beauchamp, T.L. & Childress, J.F. (1994*). Principles of Biomedical Ethics*, 4th ed. New York: Oxford University Press.

Blum, L.A. (1993). Gilligan and Kohlberg: Implications for moral theory. In M.J. Larrabee, (Ed.), *An Ethic of Care: Feminist and Interdisciplinary Perspectives* (pp. 49-68). New York: Routledge.

Buhle, M.J. & Buhle, P. (Eds.). (1978). *The Concise History of Women's Suffrage*. Urbana: University of Illinois Press.

Card, C. (1990). Caring and evil. *Hypatia, 5*, 101-108.

Chodorow, N. (1978). *The Reproduction of Mothering: Psychoanalysis and the Sociology of Gender*. Los Angeles: University of California Press.

Clouser, D. & Gert, B. (1990). A critique of principlism. *Journal of Medicine and Philosophy, 15*, 219-236.

Colby, A. & Kohlberg, L. (1987). *The Measurement of Moral Judgment*, Vol. I. *Theoretical Foundations and Research Validation*. Cambridge, London: Cambridge University Press.

Corea, G. (1985). The reproductive brothel. In G. Corea, et al. (Eds.) *Man-Made Women: How New Reproductive Technologies Affect Women* (pp. 38-51). London: Hutchinson.

Council on Ethical and Judicial Affairs, American Medical Association. (1991). Gender disparities in clinical decision making. *Journal of the American Medical Association, 266*, 559-562.

Donagan, A. (1968). Is there a credible form of utilitarianism? In M. Bayles (Ed.). *Contemporary Utilitarianism* (pp. 187-202). Garden City, New York: Doubleday and Company).

DuBose, E.R. & Hamel, R.P. (1995). Casuistry and narrative: Of what relevance to HECs? *HEC Forum, 7*, 211-227.

DuBose, E., Hamel, R., & O'Connell, L.J. (1994). *A Matter of Principles? Ferment in U.S. Bioethics*. Valley Forge, Pennsylvania: Trinity Press International.

Elshtain, J.B. (1995). The communitarian individual. In A. Etzioni, (Ed.). *New Communitarian Thinking: Persons, Virtues, Institutions, and Communities*. (pp. 99-109). Charlottesville: University of Virginia Press.

Emanuel, E.J. (1991). *The Ends of Human Life: Medical Ethics in a Liberal Polity*. Cambridge, Massachusetts: Harvard University Press.

Etzioni, A. (1998). Introduction: A matter of balance, rights, and responsibilities. In A. Etzioni, (Ed.). *The Essential Communitarian Reader* (pp. ix-xxiv). Lanham, Maryland: Rowman & Littlefield.

Fisher, S. (1986). *In the Patient's Best Interest: Women and The Politics Of Medical Decisions*. Brunswick, New Jersey: Rutgers University Press.

Fried, C. (1978). *Right and Wrong*. Cambridge, Massachusetts: Harvard University Press.

Friedan, B. (1974). *The Feminine Mystique*. New York: Dell.

Frye, M. (1991). A response to *Lesbian Ethics*: Why ethics? C.Card (ed.), *Feminist Ethics* (pp. 52-59). Lawrence, Kansas: University of Kansas Press.

Gilligan, C. (1977). Concepts of the self and of morality. *Harvard Educational Review, 47*, 481-517.

Gilligan, C. (1982). *In a Different Voice: Psychological Theory and Women's Development*. Cambridge, Massachusetts: Harvard University Press.

Green, R. (1990). Method in bioethics: A troubled assessment. *The Journal of Medicine and Philosophy, 15*, 188-189.

Hamilton, J. & Parry, B. (1983). Sex-related differences in clinical drug response: Implications for women's health. *Journal of the American Medical Women's Association, 38*, 126-132.

Hammerschmidt, D.E. & Keane, M.A. (1992). Institutional review board (IRB) review lacks impact on readability of consent forms for research. *American Journal of the Medical Sciences, 304*, 348-351.

Hoagland, S.L. (1988). *Lesbian Ethics: Toward New Values*. Palo Alto, California: Institute of Lesbian Studies.

Hoagland, S.L. (1991). Some thoughts about "Caring." In C. Card (Ed.), *Feminist Ethics* (pp. 246-263). Lawrence, Kansas: University of Kansas Press.

Hooker, W. (1849). *Physician and Patient*. New York: Baker and Scribner.

Jonsen, A.R. (1995). Casuistry: An alternative or complement to principles? *Kennedy Institute of Ethics Journal, 5*, 237-251.

Jonsen, A.R. (1991). Casuistry as methodology in clinical ethics. *Theoretical Medicine, 12*, 295-307.

Jonsen, A.R. (1986). Casuistry and clinical ethics. *Theoretical Medicine, 7*, 65-74.

Jonsen, A.R. & Toulmin, S. (1988). *The Abuse of Casuistry*. Berkeley, California: University of California Press.

Kant, I. (1956). *Groundwork of the Metaphysics of Moral*. (H.J. Paton trans.). New York: Harper & Row.

Kohlberg, L. (1976). Moral stages and moralization: The cognitive-developmental approach. In T. Lickona (Ed.), *Moral Development and Behavior: Theory, Research, and Social Issues* (pp. 31-53). New York: Holt, Rinehart, and Winston.

Kuczewski, M.G. (1994). Casuistry and its communitarian critics. *Kennedy Institute of Ethics Journal, 4*, 99-116.

Kuczewski, M.G. (1997). *Fragmentation and Consensus: Communitarian and Casuist Bioethics*. Washington, D.C.: Georgetown University Press.

Levine, R.J. (1978). Appropriate guidelines for the selection of human subjects for participation in biomedical and behavioral research. In the National Commission for the Protection of Human Subjects of Biomedical and Behavioral Research, Department of Health, Education, and Welfare. *The Belmont Report: Ethical Principles and Guidelines for the Protection of Human Subjects of Research* (Appendix I, 4-1 to 4-103). Washington, D.C.: U.S. Government Printing Office [Pub. No. (OS) 78-0013].

Manning, R.C. (1992). *Speaking from the Heart: A Feminist Perspective on Ethics.* Lanham, Maryland: Rowman & Littlefield.

Meade, C.D. & Howser, D.M. (1992). Consent forms: How to determine and improve their readability. *Oncology Nursing Forum, 19,* 1523-1528.

Merton, V. (1996). Ethical obstacles to the participation of women in biomedical research. In S.M. Wolf (Ed.). *Feminism and Bioethics: Beyond Reproduction* (pp. 216-251). New York: Oxford University Press.

Merton, V. (1993). The exclusion of pregnant, pregnable, and once-pregnable people (a.k.a. women) from biomedical research. *American Journal of Law & Medicine, 19,* 369-451.

Mill, J.S. (1911). *On the Subjection of Women.* New York: Frederick A. Stokes Company.

Mill, J.S. (1863). Utilitarianism. In J.P. Sterba (Ed.). *Ethics: The Big Questions* (pp. 119-132). Malden, Massachusetts: Blackwell Publishers, 1998.

Mitchell, J. (1974). *Psychoanalysis and Feminism.* New York: Pantheon.

Mitchell, J. (1971). *Woman's Estate.* New York: Vintage Books.

Moore v. Regents of the University of California, 793 P.2d 479 (Cal. 1990).

Nagel, T. (1986). *The View From Nowhere.* New York: Oxford University Press.

Nicholson, L.J. (1993). Women, morality, and history. In M.J. Larrabee, (Ed.). *An Ethic of Care: Feminist and Interdisciplinary Perspectives* (pp. 87-101). New York: Routledge.

Noddings, N. (1984). *Caring: A Feminine Approach to Ethics and Moral Education.* Berkeley, California: University of California Press.

Peppin, P. (1994). Power and disadvantage in medical relationships. *Texas Journal of Women and Law, 3,* 221-263.

Purdy, L.M. (1992). In H.B. Holmes & L.M. Purdy (Eds.). *Feminist Perspectives in Medical Ethics* (pp. 8-13). Bloomington, Indiana: Indiana University Press.

Raskin, A. (1974). Age-sex differences in response to antidepressant drugs. *Journal of Nervous and Mental Diseases, 159,* 120-130.

Rawls, J. (1971). *A Theory of Justice.* Cambridge, Massachusetts: Belknap Press.

Reiman, J. (1988). Utilitarianism and the informed consent requirement (or: should utilitarians be allowed on medical research ethical review boards?). In B.A. Brody (Ed.). *Moral Theory and Moral Judgments in Medical Ethics* (pp. 41-51). Boston: Kluwer Academic Publishers.

The Responsive Communitarian Platform: Rights and Responsibilities.(1998). In A. Etzioni (Ed.). *The Essential Communitarian Reader* (pp. xxv-xxxix). Lanham, Maryland: Rowman & Littlefield.

Rich, A. (1980). Compulsory heterosexuality and lesbian existence, *Signs, 5,* 631-660.

Rivera, R., Reed, J.S., & Menius, D. (1992). Evaluating the readability of informed consent forms used in contraceptive clinical trials. *International Journal of Gynecology and Obstetrics, 38,* 227-230.

Roberts, D.E. (1991, May). Punishing drug addicts who have babies: Women of color, equality, and the right of privacy. *Harvard Law Review, 104,* 1419-1482.

Rosser, S.V. (1989). Re-visioning clinical research: Gender and the ethics of experimental design. *Hypatia, 4,* 125-139.

Rosser, S. V. (1992). Re-visioning clinical research: Gender and the ethics of experimental design. In H.B. Holmes & L.M. Purdy (Eds.). *Feminist Perspectives in Medical Ethics* (pp. 128-139). Bloomington, Indiana: Indiana University Press.

Sen, A. (1975). *On Ethics and Economics.* Oxford: Basil Blackwell.

Sherwin, S. (1992a). Feminist and medical ethics: Two different approaches to contextual ethics. In H.B. Holmes & L.M. Purdy (Eds.). *Feminist Perspectives in Medical Ethics* (pp. 17-31). Bloomington, Indiana: Indiana University Press.

Sherwin, S. (1992b). *No Longer Patient: Feminist Ethics and Health Care.* Philadelphia: Temple University Press.

Sherwin, S. (1994b). Women in clinical studies: A feminist view. In A. Mastroianni, R. Faden, & D. Federman (Eds.). *Ethical and Legal Issues of Including Women in Clinical Studies,* vol. 2 (pp. 11-17). Washington, D.C.: National Academy Press.

Sichel, B.A. (1991). Different strains and strands: Feminist contributions to ethical theory. *Newsletters on Feminism and Philosophy, 90,* 86-92.

Smart, J.J. C. (1961). *An Outline of a System of Utilitarian Ethics.* Melbourne: University Press.

Smith, J.F. (1996). Communicative ethics in medicine: The physician-patient relationship. In S.M. Wolf (Ed.). *Feminism and Bioethics: Beyond Reproduction* (pp. 184-215). New York: Oxford University Press.

Stanton, E.C. (1993). *The Woman's Bible*. Boston: Northeastern University Press.

Strong, C. (1997). *Ethics in Reproductive and Perinatal Medicine*. New Haven: Yale University Press.

Todd, A.D. (1983). *Intimate Adversaries: Cultural Conflict Between Doctors and Women Patients*. Philadelphia: University of Pennsylvania Press.

Tong, R. (1993). *Feminine and Feminist Ethics*. Belmont, California: Wadsworth Publishing Company.

Tong, R. (1996). Feminist approaches to bioethics. In S.M. Wolf (ed.). *Feminism & Bioethics: Beyond Reproduction* (pp. 67-94). New York: Oxford University Press.

Tong, R. (1997). *Feminist Approaches to Bioethics: Theoretical Reflections and Practical Applications*. Boulder, Colorado: Westview Press.

Tronto, J.C. (1993). Beyond gender difference to a theory of care. In M.J. Larrabee, (Ed.). *An Ethic of Care: Feminist and Interdisciplinary Perspectives* (pp. 240-257). New York: Routledge.

Walker, M.U. (1998). *Moral Understandings: A Feminist Study in Ethics*. New York: Routledge.

Warren, V.L. (1992). Feminist directions in medical ethics. In H.B. Holmes & L.M. Purdy (Eds.). *Feminist Perspectives in Medical Ethics* (pp. 32-45). Bloomington, Indiana: Indiana University Press.

West, C. (1983). "Ask me no questions," an analysis of queries and replies in physician-patient dialogues. In S. Fisher & A.D. Todd (Eds.). *The Social Organization of Doctor-Patient Communication* (pp. 75-106). Washington, D.C.: Center for Applied Linguistics.

West, R.C. (1988). *Jurisprudence and Gender*. University of Chicago Law Review, 55, 1-72.

Whitbeck, C. (1972). The maternal instinct. In J. Treblicot, (Ed.). *Mothering: Essays in Feminist Theory* (pp. 185-192). Totowa, New Jersey: Rowman and Allanheld Publishers.

Williams, B. & Smart, J.J.C. (1973). *Utilitarianism, For and Against*. Cambridge: Cambridge University Press.

Winkler, E.R. (1993). From Kantism to contextualism: The rise and fall of the paradigm theory in bioethics. In E.R. Winkler & J.R. Coombs (Eds.). *Applied Ethics: A Reader* (pp. 343-365). Cambridge, Massachusetts: Blackwell Publishers.

Wollstonecraft, M. (1988). *A Vindication of the Rights of Women*. M. Brody (Ed.). London: Penguin Books.

3
ETHICAL ISSUES BEFORE THE STUDY BEGINS

STUDY DESIGN

Designing the Study

The study design lays the foundation for the research that will be conducted. (The Appendix provides a brief review of study designs used in research.) The study design is critical both scientifically and ethically, regardless of one's theoretical orientation. A poor study design will yield unusable results. This fails to maximize good pursuant to a utilitarian perspective. Subjecting individuals to risk, no matter how slight, where no benefit can be derived could constitute a lack of respect for persons (deontology) and a devaluation of the participants (principlism, ethic of care). The design of a study must be such that it is both scientifically optimized and "ethically optimized" (Coughlin, 1996: 145) to the extent possible. However, on occasion, the scientific optimization of a research design may be unethical and compromises must be made in the design of the study. The injunction to have a scientifically valid study design is reflected in numerous international documents. For instance, the Nuremberg Code requires that

> The experiment should be such as to yield fruitful results for the good of society, unprocurable by other methods or means of study, and not random or unnecessary in nature.

Commentary of Guideline 14 of the *International Ethical Guidelines for Biomedical Research Involving Human Subjects* (CIOMS, 1993: 38) provides that "scientifically unsound research on human subjects is *ipso facto* unethical in that it may expose subjects to risk or inconvenience to no purpose." The Helsinki Declaration requires that

> [b]iomedical research involving human subjects...conform to generally accepted scientific principles and...be based on adequately performed laboratory and animal experimentation and on a thorough knowledge of the scientific literature.

This standard enunciated in the Helsinki Declaration is incorporated by reference into the World Health Organization's (1995) *Guidelines for Good Clinical Practice (GCP) for Trials on Pharmaceutical Products.*

Drug trials often present many issues requiring resolution. For instance, a crossover design is appropriate in situations in which the disease condition is stable and, if relieved, is not permanently cured by either the experimental treatment or the comparison treatment. (See Appendix 1 for an explanation of crossover trials.) Additionally, there can be no carry over effect from the first treatment assignment to the second treatment assignment. However, it may be difficult to determine if the condition is stable. And, even if it is stable, there is a washout period between the administration of the experimental and comparison treatments, which may threaten the stability of the condition and, consequently, place the participant at increased risk.

Drug trials involving a "placebo washout" may be particularly troublesome ethically. A "placebo washout" is a period of time following consent to participate and prior to the initiation of the study when the participants are withdrawn from other drugs they may be using and receive, instead, a placebo. The administration of the placebo and the washout period help to assess whether the participants will respond to a placebo. They also provide an opportunity to eliminate the effects of any drugs that the participants may have been using prior to the commencement of the trial and to determine if the participants intend to cooperate with instructions about taking drugs (Jones and Kenward, 1989). However, if vulnerable participants are withdrawn from an effective therapy, they may suffer injury.

Study design may be an issue, however, even in studies that do not involve experimentation. Consider, for instance, the discussion of the Tearoom Trade in chapter 1. Here, the investigator deliberately misled the participants—who did not know that they were participating in a study--about the nature of his role. Advocates of the use of deception in research maintain that it produces knowledge that would not otherwise be obtainable, as was the case with the Tearoom Trade, and that critics of this technique have overestimated its dangers (Berkowitz, 1978; Baron, 1981). Some assert that the use of deception will help to reduce the Hawthorne effect, whereby research participants may alter their behavior solely because they are receiving attention in being studied (Levine and Cohen, 1974). (The biomedical counterpart to deception would be use of a placebo, discussed above in the context of the placebo washout and below in the section dealing with choice of comparison treatment. In this context, however, unlike the Tearoom Trade, study participants are informed that they are participating in research; information is withheld with respect to the substance that they are to receive.)

Baumrind (1978, 1979) has deplored the use of deception as a research strategy on ethical, psychological, societal, and scientific grounds. She has charged that deception results in the impairment of the participant's ability to endow relationships with meaning, reduces the participant's trust in authority, reduces respect for science, negatively impacts on an individual's ability to trust his or her own judgment, and may impair the individual's self-esteem. Margaret Mead (1970: 167) was highly critical of the use of deception, fearing its ultimate impact not only on the research subjects, but also on the researchers and the research:

> Besides the ethical consequences that flow from contempt for other human beings, there are other consequences—such as increased selective insensitivity or delusions of grandeur and omnipotence—that may in time seriously interfere with the very thing which he has been attempting to protect: the integrity of his own scientific work. Encouraging styles of research and intervention that involve lying to other human beings therefore tends to establish a corps of progressively calloused individuals, insulated from self-criticism and increasingly available for clients who can become outspokenly

cynical in their manipulating of other human beings, individually and in the mass.

Various commissions and international documents have attempted to address the use of deception in research. The National Commission for the Protection of Human Subjects of Biomedical and Behavioral Research (1978: 26-27) commented:

> In some research there is concern that disclosure to subjects or providing an accurate description of certain information, such as the purpose of the research or the procedures to be used, would affect the data and the validity of the research. The IRB can approve withholding or altering such information provided it determines that the incomplete disclosure or deception is not likely to be harmful in and of itself and that sufficient information will be disclosed to give subjects a fair opportunity to decide whether they want to participate in the research. The IRB should also consider whether the research could be done without incomplete disclosure or deception. If the procedures involved in the study present risk of harm or discomfort, this must always be disclosed to the subjects. In seeking consent, information should not be withheld for the purpose of eliciting the cooperation of subjects, and investigators should always give truthful answers to questions, even if this means that a prospective subject becomes unsuitable for participation. In general, where participants have been deceived in the course of research, it is desirable that they be debriefed after their participation.

Table 1 sets forth various questions that may assist the investigator in determining whether the study is ethically optimized. Institutional review boards (IRBs) are discussed below.

Exercise

You have been asked to serve as a consultant to a clinical trial that will assess whether Projoy, a new compound, is effective for the relief of mild to moderate clinical depression, a syndrome experienced by a large proportion of the U.S. population. The drug will be tested in various populations.

1. Explain what, if any, ethical issues may arise in the context of this study. Be specific.

2. Assume for the purpose of this subpart only, that you must advise whether to conduct a traditional clinical trial or a crossover trial. Explain the ethical and legal implications, if any, associated with each study design.

TABLE 1. Questions To Guide the Ethical Optimization of a Study

What harms may result from participation in the study? Consider not only health-related harms, but harms in a variety of spheres, including economic, emotional, social, and legal. Consider the potential harms from the perspective of the participant, not only the investigator.

What benefits may ensue from participation in the study? Consider not only health-related harms, but harms in a variety of spheres, including economic, emotional, social, and legal. Consider the potential harms from the perspective of the participant, not only the investigator. Do not include economic incentives in this calculation. (See chapter 4.)

Can the harms be eliminated or reduced by changing the type of study being conducted? For instance, could the harm be reduced by utilizing a classical clinical trial rather than a cross-over trial? By using historical controls rather than an actual comparison group? By utilizing a case-control design rather than a cohort design?

If a different study design is used, what are the scientific implications? If a different study design is not used, what are the scientific and ethical implications?

Formulating the Research Team

One critical component of the study design is the designation of the research team. If the research team is not competent to carry out the study protocol, the research may not be usable. This is problematic ethically regardless of one's ethical orientation because, again, the research may put participants at risk with no benefit being derived. From a utilitarian standpoint, this does not maximize good. Using the four principles as a framework, it fails to show respect for persons and may violate the principles of beneficence and nonmaleficence by subjecting participants to risk with no benefit. The feminine/feminist ethic of care may be violated because such actions fails to consider the participants' feelings with respect to a meaningless contribution of their time and energy.

Various international documents reflect the requirement that investigators be competent. Paragraph 8 of the Nuremberg Code states that

> [t]he experiment should be conducted only by scientifically qualified persons. The highest degree of skill and care should be required through all stages of the experiment of those who conduct or engage in the experiment.

The Helsinki Declaration also stipulates that

[b]iomedical research involving human subjects should be conducted only by scientifically qualified persons and under the supervision of a clinically competent medical person. The responsibility for the human subject must always rest with a medically qualified person and never rest on the subject of the research, even though the subject has given his or her consent.

The *Guidelines for Good Clinical Practice (GCP) for Trials on Pharmaceutical Products* (1995) also set forth criteria regarding the qualifications of the investigator. Paragraph 4.2 provides that the investigator should

> have qualifications and competence in accordance with local laws and regulations as evidenced by an up-to-date curriculum vitae and other credentials (decisions relating to, and the provision of, medical or dental care must always be the responsibility of a clinically competent person legally allowed to practise medicine or dentistry);
>
> have good knowledge and experience of the field of medicine or dentistry defined by the protocol;
>
> be experienced in clinical trial research methods or receive scientific support from an experienced colleague;
>
> be aware of available relevant data and literature and all information provided by the sponsor;
>
> have access to human and other resources to assume full responsibility for the proper conduct of the trial;
>
> be aware of any and comply with national regulatory and legal and ethical requirements.

Many research studies today require the collaboration of professionals from numerous disciplines due to the complexity of the research and the analysis of the results. It is not unusual, for instance, to find that a research team for a particular project comprises clinicians, epidemiologists, social scientists, biostatisticians, and information systems personnel.

Deficiencies in a research team may have serious ethical implications. First, a lack of clinical expertise in the appropriate field may lead to an inability to monitor injury or health of the participants, resulting in an increased risk to them. Lack of biostatistical and/or epidemiological expertise may result in a lack of understanding with respect to the modeling of the results, the statistical power of the study to detect changes or differences, and an inability to adequately interpret the findings. Such deficiencies may render the study results unusable, so that the participants were needlessly exposed to risk with little wider benefit resulting from the study.

Exercise

A research team received funding to conduct a clinical trial to assess the efficacy of various drug treatments against a standard treatment. Accordingly, there are five arms of the study: standard treatment, standard treatment + Drug A, standard treatment + Drug B, Drug A only, and Drug B only. The research team consists of a clinician experienced in treating the disease of interest and a pharmacist with extensive experience in drug evaluations.
1. What deficiencies in the research team are present?
2. What are the potential ethical implications of these deficiencies?

Formulating Inclusion and Exclusion Criteria

In formulating inclusion and exclusion criteria—who should be permitted to participate and who should not participate—the investigator must obviously consider the scientific goals of the study. For instance, if the primary goal of the study is to identify the risk factors for a chronic disease affecting only young children, it does not make sense to include in the study either older children or adults, who are not susceptible to the disease. Exclusion criteria are useful to restrict entry into the study to reduce analytical difficulties resulting from confounding (see appendix). However, care must be taken so as not to overly restrict participation in the study, which could raise serious scientific and ethical issues.

Consider, for instance, a situation where eligibility for clinical trials, or most clinical trials, is limited to men. In some ways, this makes sense scientifically. The more homogeneous the sample, the easier the statistical analysis. If women are included in, for instance, a clinical trial of a new drug, additional confounding may be introduced due to large weight and size differences, metabolic differences, and hormonal differences. The statistical analysis, as a result, may be significantly more complex. Consider, however, the adverse scientific consequences of failing to include women. The drug will have been tested only on men, who are different from women in size and metabolism. Consequently, dosages established for men may or may not be valid for women. The adverse effects of the drug in men may not be exactly the same as in women because of size and metabolic differences. Additionally, various contraindications for the use of the drug may not even be detected by relying on an all male sample because some drugs, for instance, that are commonly used by women, such as contraceptive pills, are never used by men (Institute of Medicine, 1994).

Ethically, the failure to include women in most clinical trials is also problematic. Utilizing a principlistic framework, we can see that the exclusion of women from clinical trials in general deprives them of sharing in the burdens and benefits of research, thereby violating the principle of justice. It could also be asserted that it reflects a lack of respect for persons because it fails to accommodate women's right to autonomy, to make their own decisions regarding participation. From a communitarian perspective, it can be argued that their exclusion is, in essence, exclusion from a community and the benefits and burdens available to that community. Feminist ethicists might argue that the exclusion of women from clinical trials is a reflection of the systemic and systematic oppression and subjugation of women and a denial of their right to participate in public space.

Concern regarding the inclusion of diverse groups is reflected in Guideline 10 of the *International Ethical Guidelines for Biomedical Research Involving Human Subjects* (CIOMS, 1993: 29):

> Individuals or communities to be invited to be subjects of research should be selected in such a way that the burdens and benefits of the research will be equitably distributed. Special justification is required for inviting vulnerable individuals and, if they are selected, the means of protecting their rights and welfare must be particularly strictly applied.

Recruitment from diverse groups is discussed in chapter 4. Who may be considered to be a vulnerable individual is discussed below. Special protections for vulnerable individuals are discussed in the context of informed consent in chapter 4.

Exercise

Assume for the purpose of this questions that you are testing Projoy as indicated in the previous exercise. The drug has not been found to be safe or unsafe in pregnant women.

1. Utilizing a principlistic framework, explain the ethical issues specific to the enrollment of pregnant women for this clinical trial and how you would resolve them.

2. Utilizing a communitarian perspective, explain the ethical issues specific to the enrollment of pregnant women for this clinical trial and how you would resolve them.

Working with Vulnerable Participants

A researcher must address the appropriateness of conducting research with a vulnerable population, and the manner in which that research is conducted, regardless of his or her adherence to a particular ethical orientation. For instance, reliance on deontological thought requires that persons participating in research be respected. That necessarily requires an examination of what constitutes respect and how that can be effectuated when working with individuals who may be especially vulnerable, such as those with diminished capacity to make decisions. A utilitarian perspective demands an examination of whether the greatest good is served by the inclusion or exclusion of vulnerable individuals from research. Although a cursory analysis of the issue might seem to favor the inclusion of vulnerable populations due to the apparent ease of obtaining cooperation, a more thorough investigation raises additional subissues that require attention. For instance, if vulnerable populations participate in research, without the provision of special protections, what is ultimately the effect of such research on public confidence in science? Feminist thought, with its emphasis on caring and relationships, might ask how and whether research involving vulnerable participants can be conducted in a manner that promotes nonviolence and is sensitive to the needs of the participants.

Who Is Vulnerable?

First and foremost, a determination must be made as to which persons or groups are potentially vulnerable. The Nuremberg Code offers some guidance on this in its requirement of informed consent:

> The voluntary consent of the human subject is absolutely essential.
> This means that the person involved should have legal capacity to give consent; should be so situated as to be able to exercise free power of choice, without the intervention of any element of force, fraud, deceit, duress, overreaching, or other ulterior form of constraint or coercion; and should have sufficient knowledge and comprehension of the elements of the subject matter involved as to enable him to make an understanding and enlightened decision. This latter element requires that before the acceptance of an affirmative decision by the experimental subject there should be made known to him the nature, duration, and purpose of the experiment; the method and means by which it is to be conducted; all inconveniences and hazards reasonably to be expected; and the effects upon his health or person which may possibly come from his participation in the experiment.

This tells us, then, that individuals who may be vulnerable are those who do not have legal capacity to consent, those who are unable to exercise free choice, and those who do not have sufficient knowledge or understanding to be able to make a choice.

Some individuals might argue that the Nuremberg Code had no relevance to nations that did not participate in its formulation, such as countries that, at the time of World War II, were still possessions of foreign powers. Others might assert that who may be considered to be vulnerable in one society may not be so in another. These arguments reflect the division between universalists and relativists that we considered in chapter 1 (Beauchamp, 1996). Stated somewhat simplistically, universalists maintain that there is a set of principles that apply to everyone and that govern all, regardless of the society or culture in which they function. Relativists assert that the validity of specific values varies between cultures (Beauchamp, 1996). Levine (1996) has proposed a compromise between these two stances, which would recognize broad, universally acceptable principles, but would also recognize the existence and legitimacy of variation in interpretation and application across cultures and societies. Using Levine's formulation, it would seem that there would be general agreement on the broad definition of who may be vulnerable, although the classes of individuals encompassed by that definition may vary. Accordingly, we see that the *International Ethical Guidelines for Biomedical Research Involving Human Subjects*, promulgated by the Council for International Organizations of Medical Sciences and the World Health Organization (1993), has delineated several groups that could be considered to be vulnerable, including prisoners, children, and individuals with mental or behavioral disorders. Too, various societies that have developed formal regulations for research involving humans have delineated different groups. For instance, the United States' regulations enumerate children, prisoners, and women who are pregnant or who are able to become pregnant (45 Code of Federal Regulations sections 46.201-.211, .301-.306, .401-.409, 1999). Uganda's enumeration of categories of persons who are especially vulnerable is different, and derive from Uganda's history and experience: soldiers, prisoners, refugees, children, and individuals with mental or behavioral disorders (Loue, Okello, and Kawuma, 1996).

The question remains, however, how to determine whether an individual or class of individuals meets the criteria enumerated above to warrant classification as vulnerable. Some such determinations will not be left to the judgment of the investigator, such as in situations where the law specifies that certain classes of individuals are to be considered vulnerable. For instance, United States regulations dictate that children are to be considered vulnerable in the context of research, regardless of their age or mental maturity. In other situations, the investigator may need to look beyond the regulations that delineate this status; because a class of individuals is not considered by law to be vulnerable in the context of research, does not foreclose the possibility that ethically the class may be especially vulnerable. This judgment requires a more in-depth examination of the criteria enunciated above.

<u>Lack of legal capacity to consent</u>. Whether an individual lacks the legal capacity to consent is determined by reference to the controlling law. For instance, in the United States, minors are generally considered to lack legal capacity to consent. There are, however, certain exceptions, depending upon the relevant state law. These may include minors who have been emancipated through a judicial finding of such, minors who have married or have become pregnant, or, in the context of clinical care, minors who present for the diagnosis and/or treatment of a sexually transmitted disease.

A second category of individuals often deemed to lack legal capacity to consent consists of some individuals who have been declared to be incompetent by a court. A distinction must first be made between competence and capacity. These terms are often used interchangeably, but they actually refer to different concepts. All adults are presumed to be legally competent. A determination of incompetence must be made by a judge in accordance with legally mandated procedures and standards. Although a judicial finding that an individual is incompetent does not mean that the individual necessarily lacks the ability to make decisions for him- or herself with respect to specified matters, many courts do not make specific findings with respect to individuals' abilities in specified areas. Consequently, the court may delegate to a conservator or guardian the authority to make all decisions for the incompetent individual. Where all such authority has been delegated, the individual may lack legal capacity to consent, although he or she may, in actuality, have capacity to do so (Loue, in press).

Some persons may lack actual capacity to consent, but have not been found to lack legal capacity to consent, i.e., have not been found to be incompetent. This includes, for instance, those who may be permanently or temporarily cognitively impaired due to the effects of a particular disease or condition. For instance, an individual suffering from schizophrenia may not have been adjudicated as incompetent, but his or her ability to understand may fluctuate. Such situations are discussed below, in connection with inadequate understanding, and in chapter 4.

Inability to exercise free choice. Certain situations render it more unlikely that an individual will feel free to make a choice regarding his or her participation in a study. For instance, where an individual is confined and is subject to the supervision and discretion of those in charge of that confinement, it may be difficult to refuse to participate due to fear of either possible repercussions, such as physical punishment or of the withholding of benefits, such as medical care or earlier release from prison. For this reason, some believe that it is not possible for anyone in an institutional or conscripted situation to exercise truly free choice (see Goffman, 1961). Other situations may appear to allow the possibility of free choice, but further examination may indicate that this is not the case. For instance, a village leader may say that individuals in the village have the ability to decide for themselves to participate or not, but the reality may be that, once the leader has approved the study or program, individuals who are found to have refused to participate may suffer subtle or blatant forms of political or economic punishment for exercising that choice.

Inadequate understanding. This criterion may be the most difficult of the three in application. In some situations, it is relatively easy to ascertain that the prospective participant is not capable of understanding the study or his or her role in the study. This could occur, for instance, where the prospective participant is an infant or toddler, or where the "participant" is wheeled into the emergency department of a hospital, unconscious. Other situations are considerably more complex, however. For instance, if an individual is schizophrenic and has considerable difficulty understanding the consequences of a particular action, although he or she may be able to recite back what is said, it may be unclear how much is truly understood. Some researchers have suggested performing short mental status examinations as a means of assessing understanding in the clinical setting (Lo, 1990). However, mini-mental examinations often assess an individual's orientation to the here and now. They have often been criticized as being culturally biased. Additionally, they may reflect an individual's ability to memorize everyday facts, such as the name of the president, but do not reflect an individual's awareness of the implications of those facts.

These properties—legal capacity, free choice or voluntariness, and understanding—are critical in the formulation of informed consent in the context of a particular study and in the evaluation of whether an individual gave informed consent. They are discussed in greater detail in this context in chapter 4.

Why Involve Vulnerable Participants?

The second question that must be asked is why the proposed research is being conducted with vulnerable participants and, if so, whether their involvement is necessary. This question is critical. As seen in chapter 1, many research studies in the past have relied on vulnerable subjects because their participation was relatively more convenient and economical, and not because the research benefited them as a class or because it would result in knowledge relevant to a particular condition that they might have had, such as epilepsy or mental illness.

There may be valid scientific reasons to conduct a study involving vulnerable participants. For instance, major organs change in size and function between infancy and childhood and adulthood. A procedure or treatment or dosage that may be effective during one stage of growth may have to be modified during another phase of growth. It cannot be assumed that what is effective or even safe for an adult will also be effective and safe for a child or infant (Kaufman, 1994). Some conditions may affect primarily certain groups, or only certain groups, so that if research is not conducted with these specific groups of individuals, it will be difficult to learn more about the condition or effective and safe treatments for the condition. This is the case, for instance, with newborn respiratory distress syndrome, which is unique to newborns, and neuroblastoma, which occurs in young children (Kaufman, 1994). It would be difficult, if not impossible, to assess the effectiveness of a new treatment for schizophrenia without involving schizophrenic individuals in the research.

There are also numerous ethical bases for the involvement of vulnerable participants in the research. First, the study may offer a potential benefit to the participant. For instance, if a mentally ill individual is participating in a clinical trial to evaluate the efficacy of a new drug which, it is believed, may reduce symptoms of paranoia, the individual may derive a health benefit, despite the randomization procedure, because he will receive either the experimental drug or the comparison drug. (See appendix for a discussion of clinical trials.) Even where the individual does not receive a direct benefit, he or she may derive satisfaction from knowing that he or she is participating in research that may ultimately lead to a remedy for others. Additionally, the participation may be justified because it will help others.

Whether any of the reasons proffered above provides a sufficient basis for the involvement of vulnerable persons in research may depend on one's ethical perspective. For instance, utilitarians may argue that vulnerable persons should be permitted to participate in research because, ultimately, their participation will result in the maximization of good because it will produce knowledge that will help us to reduce symptoms or cure or prevent disease. Communitarians might argue that, in balancing individual rights with social responsibility, individuals who need additional protection have the right and the responsibility to participate in research where it is anticipated that the research will yield benefits to the larger community. Additionally, the community then, might be said to have the responsibility to permit that participation and to provide additional protections, as necessary, to facilitate its occurrence. Those subscribing to distributive justice, as formulated by Rawls, might assert that vulnerable participants must be permitted to share in the benefits and burdens of research. However, in view of the nonexistence of "equal right to the most extensive liberty compatible with a similar liberty for others" in reality, it could be argued that vulnerable participants must somehow be safeguarded in the process of participation.

International documents reflect the difficulty of involving vulnerable participants in research. The Nuremberg Code essentially excludes such individuals from participation because it requires that each individual provide his or her own voluntary consent to participate. If the individual lacks the capacity to consent, then clearly he or she cannot participate. The *International Ethical Guidelines for Biomedical Research Involving Human Subjects* (CIOMS and WHO, 1993: Commentary to Guideline 10, 30) furnishes the following guidance in deciding whether to include vulnerable participants:

> Ethical justification of [the involvement of vulnerable individuals or groups] usually requires that investigators satisfy ethical review committees that:
> --the research could not be carried out reasonably well with less vulnerable subjects;
> --the research is intended to obtain knowledge that will lead to improved diagnosis, prevention or treatment of diseases of other health problems characteristic of or unique to the vulnerable class, either the actual subjects or other similarly situated members of the vulnerable class;
> --research subjects and other members of the vulnerable class from which subjects are recruited will ordinarily be assured reasonable access to any diagnostic, preventive or therapeutic products that will become available as a consequence of the research;
> --the risks attached to research that is not intended to benefit individual subjects will be minimal, unless an ethical review committee authorizes a slight increase above minimal risk . . . ; and
> --when the prospective subjects are either incompetent or otherwise substantially unable to give informed consent, their agreement will be supplemented by the proxy consent of their legal guardians or other duly authorized representatives.

(The concept of informed consent is discussed in chapter 4.)

Guideline 5 of the same document, which is specific to reliance on children as research subjects, requires that research not be carried out with children that could be carried out equally well with adults and that the research must be designed to obtain knowledge that is relevant to the health needs of children (CIOMS and WHO, 1993: 20). Guideline 6, which addresses research involving persons with mental and behavioral disorders, similarly prohibits the conduct of such research if it can be conducted with persons who are in "full possession of their mental faculties" (CIOMS and WHO, 1993: 22).

Exercise

Because of your great expertise in research ethics and study design, the state health department of the state of Woeisme has asked you to work on a study that is scheduled to begin during the next few months.

The original study was to determine what, if any, beneficial and/or adverse effects are associated with an adolescent having or not having an abortion. The study was to utilize a prospective cohort design. It would follow all pregnant girls ages 13-19 presenting at clinics and hospitals in five major cities in the state. Girls would be asked to participate in the study after they had decided how they were going to handle their pregnancy, e.g abortion, keep the baby, give the baby up for adoption, have the baby but give him/her to a relative for care, etc. Independent variables of interest included age, level of education, income level and educational level of family members, race, ethnicity, and substance use. Baseline measurements were to be taken on the physical and mental health status of each of the participants. Outcome variables were to include level of education attained or number of years of school during the 5 year study period, physical health status, mental health status, employment status, and income level, among others.

Just prior to the initiation of the study, the state legislature passed a new infanticide law which prohibits the killing or attempted killing of a complete child. Assistance with the killing or attempted killing, or the commitment of a substantial act towards the killing or attempted killing, is also prohibited. The law is being interpreted to prohibit the killing of a fetus after 6 weeks of gestation, when a human form is discernible. Although it is unclear what constitutes "a substantial act," it is believed that prosecutors are interpreting the statute broadly so that even the provision of information regarding abortion procedures or assistance with completion of the requisite insurance forms may constitute "a substantial act," leaving the individual open to criminal prosecution. Significant criminal penalties result for anyone found convicted under this law. Health care professionals may, if convicted, lose their licenses due to the felony conviction.

1. What ethical problems now face the state public health department researchers as a result of the passage of this new law? Be sure to enumerate and describe all such problems.

2. Can the study be redesigned to avoid any or all of the problems that you noted above? If so, explain how you would design it, which problems would be resolved, and how they would be resolved. Indicate which, if any, of the problems that you noted are not resolvable with your proposed new design. If you believe that the study cannot be redesigned to avoid any of these problems, state so and support your answer.

Selecting a Comparison Treatment or Intervention

Recent research conducted in the area of HIV prevention has sparked significant controversy regarding the selection of the comparison treatment or intervention (Angell, 1997; Bayer, 1998; Cohen, 1997; Gambia Government/Medical Research

Council Joint Ethical Committee, 1998; Lurie and Wolfe, 1997; Varmus and Satcher, 1997). The question of what comparison treatment or intervention should be utilized is of concern ethically because the choice of an inappropriate comparison treatment or intervention would potentially subject research participants to greater risk than would be necessary to successfully conduct the research.

Consider, for instance, an extremely complex situation in which an investigator wishes to compare a new drug regimen to a placebo. (See the appendix for a brief discussion of clinical trials.) Assume further that a treatment has been proven to be effective, but it would not be available to individuals outside of a clinical trial in the geographic area in which the trial is being conducted, due to costs and the lack of an infrastructure required for the use of the drug, such as a reliable source of electricity for refrigeration, the availability of lab equipment and trained laboratory and medical personnel to periodically test blood levels of the drug, and a transportation system for transport to medical centers and other locations to receive the treatment and the necessary examinations. Reliance on a placebo makes the conduct of the study logistically easier and less expensive and the statistical analysis less complex. Conducting the study with the alternate treatment, rather than the placebo, could mean that the study cannot be conducted because of the immensity of the logistical and economic problems. There is also the possibility that use of the comparison drug, rather than a placebo, will induce drug resistance and create resistant strains of the infection because the drug will not be available to trial participants after the conclusion of the trial, for the reasons stated above. Regardless of whether the comparison treatment or the placebo is used with the controls, the experimental regimen has the potential to prevent disease transmission in a significant number of participants. Use of the placebo, however, may result in transmission to more participants during the course of the trial than would use of the comparison drug. There is no known cure for this disease. This description closely mirrors the situation in which controversy arose regarding the use of placebo in trials conducted in developing countries to reduce the rate of HIV transmission from pregnant mothers to their infants.

From a deontological perspective, reliance on subjects' participation without regard to their increased risk as a result of the choice of the comparison treatment could be seen as using the participants as a means, without regard to the end. Utilitarians might argue that it is better to conduct the trial using the placebo because there can be no good accomplished without the trial, which could be cancelled if the comparison treatment must be used. If the placebo is used, some individuals may be infected with the disease, but the greatest good will be effectuated by conducting the trial and ultimately learning how to prevent the transmission of the disease among a large number of persons who might otherwise become infected. Others, however, might argue that the greatest good comes about through the use of the comparison treatment because the general public will lose faith in science and medicine if they come to believe that their interests will be sacrificed for the good of others; ultimately, no one would be willing to participate in a trial so that medical knowledge would not advance.

Various conclusions could also result using a principlistic approach. The principles of nonmaleficence and beneficence could be used to support the position that the comparison treatment must be used, rather than a placebo, because to do otherwise would not be for the benefit of others and would create, rather than prevent, harm. However, if the principle of autonomy were to take precedence, it could be argued that as long as individuals were aware of the potential risks involved in being randomized to a placebo, and were willing to assume those risks, that they should be allowed to do so. To prevent them from doing so is to demonstrate a lack of respect for persons and their right to choose freely. Others might argue, however, that there is no free choice here because individuals who may transmit an incurable disease to their offspring have no real choice to begin with, since the trial represents the only potential mechanism by which to prevent that transmission. Casuists would try to analyze the situation by determining which of the facts are the most salient to the resolution of the dilemma, such as the incurability of the disease and the lack of the comparison drug's availability outside of the trial, among others. They would then use these facts, in conjunction with facts and principles derived from similar past factual situations, to arrive at an appropriate resolution.

Several of the international documents provide some guidance with respect to such situations. However, they do not, by any means, provide a resolution. The Helsinki Declaration provides with respect to medical research combined with professional care (clinical research) that:

> In any medical study, every patient—including those of a control group, if any—should be assured of the best proven diagnostic and therapeutic method.
>
> The physician can combine medical research with professional care, the objective being the acquisition of new medical knowledge, only to the extent that medical research is justified by its potential diagnostic or therapeutic value for the patient.

The *International Guidelines for Ethical Review of Epidemiological Studies* (CIOMS, 1991: guideline 44, at 21) provides that:

> Epidemiological studies that require control (comparison) or placebo-treated (i.e., non-treated) groups are governed by the same ethical standards as those that apply to clinical trials. Important principles are that:
> (i) the control group in a study of a condition that can cause death, disability or serious distress should receive the most appropriate current established therapy; and
> (ii) if a procedure being tested against controls is demonstrated to be superior, it should be offered to members of the control group.

What the Helsinki Declaration fails to make clear, however, is whether "the best proven diagnostic and therapeutic method" is one that is available in the locale in which the study is to be conducted, or one that is available at any location, at any cost, in the world. Similarly, the *International Guidelines for the Ethical Review of Epidemiological Studies* do not specify whether "the most appropriate current established therapy" is that which exists anywhere in the world, or that which exists in the locality of the trial. Neither document provides guidance as to which "best method" to use where there is divergence of opinion across countries as to the most effective treatment protocol. For instance, guidelines for the treatment of various mental illnesses differ greatly between the United States and Great Britain; what criteria should be applied to determine which is "the most appropriate" treatment in a third country that is quite dissimilar with respect to its history, health resources, health systems infrastructure, and public health profile?

This issue continues to be unresolved and respected ethicists disagree on how this provision is to be applied. Several writers have suggested scientific alternatives to this dilemma, including a placebo-controlled trial only if there is no known effective therapy that can be used as a comparison, or the participants cannot tolerate a known effective therapy, or if the treatment is so scarce that only a limited number of participants can receive it (Levine, Dubler, and Levine, 1991). The use of a lottery to decide which participants should receive the experimental treatment has been suggested in such situations (Levine, Dubler, and Levine, 1991; Macklin and Friedland, 1986).

Exercise

Assume that you have been asked by the government of Schizovia to collaborate on a trial of Projoy in that country. Schizovia appears to suffer from an extraordinarily high rate of depression in comparison with both developed countries and other underdeveloped countries, although the reasons for this high prevalence are unknown. You know from phase I and phase II trials conducted in the United States only that the serious side effects associated with other classes of antidepressant drugs are significantly less likely to occur with the use of Projoy. Further, Projoy will most likely be marketed at a significantly lesser cost than are currently available antidepressants, most of which are unavailable and unaffordable to the citizens of Schizovia. Your colleagues in Schizovia have decided to conduct a clinical trial there to evaluate the efficacy of Projoy as compared to a placebo. Discuss the ethical issues that arise as a result of this decision to use placebo in the comparison group.

Balancing the Risks and Benefits

The protocol for the design of a study should consider the balance of the risks and the benefits to the research participants. To do so reflects a manner of conduct towards the participants that considers the means and the ends (deontology). Under a principlistic framework, this approach would maximize the principles of beneficence and nonmaleficence, promoting good and preventing harm. From a utilitarian perspective, the minimization of harm would accomplish the greatest good by first, producing the least amount of harm to the fewest number of people and, second, by increasing public trust in science and research through the minimization of potential harm.

This concern for a balance of the risks and benefits to participants is reflected in various international documents governing research. The Nuremberg Code indicates that

> [n]o experiment should be conducted where there is an *a priori* reason to believe that death or disabling injury will occur; except, perhaps, in those experiments where the experimental physicians also serve as subjects.
>
> The degree of risk to be taken should never exceed that determined by the humanitarian importance of the problem to be solved by the experiment.
>
> Proper preparations should be made and adequate facilities provided to protect the experimental subject against even remote possibilities of injury, disability, or death.
>
> During the course of the experiment the human subject should be at liberty to bring the experiment to an end if he has reached the physical or mental state where continuation of the experiment seems to him to be impossible.
>
> During the course of the experiment, the scientist in charge must be prepared to terminate the experiment at any stage, if he has probable cause to believe, in the exercise of good faith, superior skill, and careful judgment required of him, that a continuation of the experiment is likely to result in injury, disability, or death to the experimental subject.

The Helsinki Declaration provides that

> [b]iomedical research involving human subjects cannot legitimately be carried out unless the importance of the objective is in proportion to the inherent risk to the subject.
>
> Every biomedical research project involving human subjects should be preceded by careful assessment of predictable risks in comparison with foreseeable benefits to the subject or to others. Concern for the interests of the subject must always prevail over the interests of science and society.

With respect to medical research combined with professional care (clinical research), the Helsinki Declaration requires that the

> potential benefits, hazards and discomforts of a new method should be weighed against the advantages of the best current diagnostic and therapeutic methods.

In the context of non-therapeutic biomedical research involving human subjects (non-clinical biomedical research), the Helsinki Declaration makes clear that:

> [I]n research on man, the interest of science and society should never take precedence over considerations related to the well-being of the subject.

All too often, however, investigators may limit their considerations of the risks and benefits to those that are related to the participant's health or medical care, and may fail to consider risks and benefits that may attend participation in the study outside of this narrow realm. For instance, potential participants could suffer economic loss, stigmatization, or loss of entitlement to other benefits as a result of their participation in or association with the study. *The International Guidelines for Ethical Review of Epidemiological Studies* (CIOMS, 1991) speaks to these circumstances:

> Investigators planning studies will recognize the risk of doing harm, in the sense of bringing disadvantage, and of doing wrong, in the sense of transgressing values. Harm may occur, for instance, when scarce health personnel are diverted from their routine duties to serve the needs of a study, or when, unknown to a community, its health-care priorities are changed. It is wrong to regard members of communities as only impersonal material for study, even if they are not harmed. (Guideline 18, at 15).

> Ethical review must always assess the risk of subjects or groups suffering stigmatization, or prejudice, loss of prestige or self-esteem, or economic loss as a result of taking part in a study. Investigators will inform ethical review committees and prospective subjects of perceived risks, and of proposals to prevent or mitigate them. Investigators must be able to demonstrate that the benefits outweigh the risks for both individuals and groups. There should be a thorough analysis to determine who would be at risk and who would benefit from the study. It is unethical to expose persons to avoidable risks disproportionate to the expected benefits, or to permit a known risk to remain if it can be avoided or at least minimized. (Guideline 19, at 15)

> Epidemiological studies may inadvertently expose groups as well as individuals to harm, such as economic loss, stigmatization, blame, or withdrawal of services. Investigators who find sensitive information that may put a group at risk of adverse criticism or treatment should be discreet in communicating and explaining their findings. When the location or circumstances of a study are important to understanding the results, the investigators will explain by what means they propose to protect the people from harm or disadvantage; such means include provisions for

confidentiality and the use of language that does not imply moral criticism of subjects' behaviour. (Guideline 21, at 16)

Similarly, the assessment of benefits should not be limited to consideration of a medical benefit inuring only to the research participant. For instance, even where a research participant receives no direct medical benefit, he or she may drive satisfaction from the knowledge that his/her participation is helping to benefit society as a whole through the addition of new knowledge. Direct payments of other forms of remuneration should not, however, be considered in an assessment of the benefits (see Macklin, 1989).

Exercise

Investigators wish to conduct a phase I clinical trial of a new form of chemotherapy for a specified form of cancer. The cancer is a rapid progressing one and, to date, the average time from diagnosis to death is six months. The cancer itself is quite painful and the patient's condition deteriorates rapidly. The only known treatment results in an extension of the patient's life for approximately six months, with a decrease in the quality of life due to the severe side effects. The new chemotherapeutic treatment is likely to be associated with severe nausea and vomiting and loss of hair, as well as a decrease in white blood cells. It is unlikely that participants in the trial will receive any direct benefit, but their participation will ultimately help in the establishment of appropriate dosing levels. How should the risks and benefits of participation be assessed and balanced? Should the trial be permitted to proceed? Why or why not?

ETHICAL REVIEW COMMITTEES

The Purpose and Function of the Review Committees

Several of the international documents developed to provide ethical guidance in the conduct of research provide for the establishment of review committees to protect the rights of research participants. For instance, consider the following.

> The primary functions of ethical review are to protect human subjects against risks of harm or wrong, and to facilitate beneficial studies. Scientific and ethical review cannot be considered separately: a study that is scientifically unsound is unethical in exposing subjects to risk or inconvenience and achieving no benefit in knowledge. Normally, therefore, ethical review committees consider both scientific and ethical aspects. (CIOMS, 1991: Guideline 40, at 20)

> The role of the ethics committee (or other board responsible for reviewing the trial) is to ensure the protection of the rights and

welfare of human subjects participating in clinical trials. (WHO, 1995, paragraph 3.2, at 10)

One might query whether the establishment of an ethical review committee is sufficient to ensure the protection of research participants' rights during the course of a study. After all, the Tuskegee study was reviewed several times and at numerous levels of government and was still permitted to proceed. One recent review of 30 research projects approved by a local ethics review committee outside of the United States found that the committee had not conducted any follow-up monitoring of the approved projects, that in approximately one-quarter of the studies informed consent forms were not completed, and in two projects there had been a change in principal investigator (Smith, Moore, and Tunstall-Pedoe, 1997). Clearly, the answer is that an ethical review committee is not sufficient to ensure that the participants' rights will be protected. However, numerous restrictions relating to participation on the review committee itself and the procedures utilized by that committee enhance the likelihood that the committee will adequately fulfill its charge.

The International Guidelines for Ethical Review of Epidemiological Studies (CIOMS, 1991) requires that ethical review committees reflect the community to be studied, that the committee members consider both personal and social perspectives in their review, and that they refrain from any unethical conduct themselves (Guidelines 37-39). The *Guidelines for Good Clinical Practice for Trials on Pharmaceutical Products* (WHO, 1995) require that the ethics committee be constituted so as to be free of the influence of anyone conducting the drug trial and that the committee develop and maintain documented policies and procedures for its work (Guideline 3.2, at 109). The ethics committee is specifically charged with the responsibility of evaluating the acceptability of the investigator to conduct the trial, the suitability of the trial, the means by which trial participants will be recruited, the adequacy and completeness of the information to be provided to potential participants, the provision of any compensation or treatment for injury or death arising in conjunction with study participation, and the appropriateness of the extent and form of payment to be made to the organizations or investigators conducting the trial and the trial participants (Guideline 3.2, at 109).

The *International Ethical Guidelines for Biomedical Research Involving Human Subjects* (CIOMS and WHO, 1993) also set forth specific criteria for the conduct of an ethical review committee. Guideline 14 requires that:

> All proposals to conduct research involving human subjects [be] submitted for review and approval to one or more independent ethical and scientific review committees. The investigator must obtain such approval of the proposal to conduct research before the research is begun.

Commentary to the guidelines provides that local review committees be organized so as to be able to provide a complete review of the proposals submitted to them, that membership include both men and women, that membership be rotated periodically, that any member of the committee with a direct interest in a specific protocol be excluded from the review of that protocol, and that specified information be provided to committee members to allow them to review the submitted protocol fully and adequately. Accordingly, investigators are required to provide a statement of the research objectives, a description of all proposed interventions, a description of any procedures to be used to withdraw to withhold standard therapies, plans for the statistical analysis of the findings, information relating to any economic or non-economic incentives or inducements to participate, information pertaining to the safety of each proposed intervention, a statement with respect to the risks and benefits of participation, a description of the mechanism to obtain informed consent, the identification of the sponsor of the research, a description of the mechanism to inform participants about harms and benefits during the study and the results of the study, an explanation of the inclusion and exclusion criteria for participation in the study, evidence of the investigator's qualifications and the existence of adequate and safe facilities in which to conduct the research, and a statement regarding the provisions made to protect confidentiality.

The *International Guidelines for Ethical Review of Epidemiological Studies* (CIOMS, 1991: Guideline 53, at 24-25) also delineates various requirements of the investigator seeking to have his or her protocol reviewed by an ethics committee. The investigator is required to submit a statement of the research objectives, a description of all proposed procedures and interventions, a plan for statistical analysis of the data, criteria for the termination of the study, and inclusion and exclusion criteria. Additionally, the protocol must set forth information relating to the safety of each proposed procedure and intervention, the mechanism for obtaining informed consent, evidence of the investigator's credentials to conduct the research, and the mechanisms to be used to safeguard confidentiality of the data.

The evaluation of the risks and benefits may be particularly problematic. Members of a review committee may subscribe to the view that healthy volunteers are able to make a free choice and can judge for themselves the acceptability of the risks against the potential benefits. According to this view, the ethical review committee should not superimpose its judgments, such as the appropriateness of the incentive, on the participants. The contrary view holds that it is the responsibility of the ethical review committee to protect participants from inducements that could impact on their ability to make a voluntary, informed choice.

Various countries have implemented procedures for both the formation and maintenance of ethical review committees and for the review of protocols submitted to those committees. A brief description is provided below of such committees and their procedures in the United States, the Netherlands, and Uganda.

Ethical Review in the United States: The Institutional Review Board

Reliance on the review of a research protocol involving human experimentation is a legal requirement under United States regulatory law. (See Appendix 2 for a discussion of regulatory law.) For instance, the Food and Drug Administration (FDA) and the Department of Health and Human Services (HHS) have each promulgated regulations governing the composition, function, and procedures of IRBs. Except where otherwise specified, this discussion addresses these institutional review committees, known in the United States as institutional review boards or IRBs, in general.

The IRB is charged with the responsibility of reviewing all proposed research to be undertaken by or at an institution to determine whether the research participants will be placed at risk. If the research involves risk, the IRB must evaluate whether the potential risks to the participant are outweighed by the potential benefit to the participant; whether the rights and welfare of the participants will be protected adequately; and whether legally effective informed consent will be obtained in a manner that is both adequate and appropriate. The IRB may also require that the researcher provide additional information to the prospective research participant, including the possibility of unforeseeable risks, circumstances in which the individual's participation may be terminated without his or her consent, costs to the participant of participating in the research, and the potential impact of the participant's decision to withdraw from the research. The IRB may also review the procedures that have been implemented to protect participants' privacy and the confidentiality of the data. Table 2 provides a summary of items to be addressed in the preparation of an IRB submission.

An IRB must adopt and follow written procedures for various functions. These include (1) the initial and continuing review of the research; (2) the reporting of its findings and actions to the principal investigator and the research institution; (3) the determination of which projects require review more frequently than once a year; (4) the determination of which projects require verification from persons other than the investigators to the effect that no material changes have occurred since the last IRB review; (5) the prompt reporting of proposed changes in the research activity to the IRB; (6) the prompt reporting to the IRB and officials of the research institution of any serious or continuing noncompliance by the investigator with the requirements or determinations of the IRB; and (7) the review of research involving children.

Table 2. Items to Be Addressed in the Preparation of an IRB Submission

Has the study design been ethically optimized?

What are the risks and benefits of participation from the vantage points of the researcher and the participant, and what is their balance?

Have protections of the participants been maximized to the greatest extent possible? Are protections for confidentiality and privacy maximized? (See chapters 4 and 5.) Are the procedures adequate to address adverse events that may arise in connection with participation?

If incentives are provided, are they proportional to participants' activities? (See chapter 4, Recruitment.)

Is the informed consent process ethically adequate? (See chapter 4.)
Does the informed consent process conform to federal and state regulations, as appropriate? (See chapter 4.)

The IRB must maintain written documentation of its activities and meetings, where applicable. This documentation should include copies of all research proposals that have been reviewed; scientific evaluations accompanying the proposals; proposed sample consent forms; investigators' progress reports; reports of injuries to research participants; copies of all correspondence between the IRB and investigators; minutes of IRB meetings, including a list of attendees and the votes on each item; a list of IRB members and their credentials; and a copy of the IRB's written procedures.

There are several major differences between the regulations of the FDA and of HHS with respect to IRBs. Many of these relate to the differences in their scope of authority and responsibility, which is discussed more fully in chapter 5. For instance, HHS permits an IRB to waive the requirement of written informed consent in research where the principal risk is a breach of confidentiality. The FDA does not include such a waiver provision because the FDA does not regulate studies which it believes would fall into that category of research. The FDA does not require that an IRB report to it changes in its membership, whereas HHS does. The FDA permits a waiver of the informed consent requirement in emergency situations, whereas HHS does not permit such a waiver (Food and Drug Administration, 1998).

A recent study found that, despite variations in the committee structure and representation of U.S. hospital-based IRBs, procedures governing research are similar (Jones, White, Pool, and Dougherty, 1996). Of 488 hospitals responding to a survey, 447 (91.6%) had an IRB. Committees had an average of 14 members, encompassing 27 medical specialties. Orthopedics was the medical specialty least likely to be represented on the IRBs (10% of the committees), followed by emergency medicine (12%) and ophthalmology (15%). In general, a proposal submitted for IRB review would go through one or more of the following steps: a critique by an IRB member prior to the committee meeting (95% of committees), subcommittee review of the proposal (35%), presentation at an IRB meeting by the investigator (69%), recommendation/vote by committee (98%), and/or final written notification or a request for revisions (91%). The most common reasons for rejecting a proposal included improperly designed consent forms (54%), poor study design (44%), unacceptable risk to research participants (34%), other legal or ethical reasons (24%), and a lack of scientific merit (14%). Few IRBs had dealt with scientific misconduct investigations (17%), and slightly more than half had a written policy relating to research integrity (58%) (Jones, White, Pool, and Dougherty, 1996).

Despite the laudatory goals underlying the formulation and implementation of IRBs, IRBs have been criticized for being overly permissive in their approval of proposed research (Classen, 1986). It is critical to recognize that approval of research by an IRB neither means that the research is ethical nor legal. Examination of the case of *Kaimowitz v. Department of Mental Health for the State of Michigan* (1973), which involved research approved by a Scientific Review Committee, is instructional. This case involved an action filed by Kaimowitz on behalf of an unnamed individual (John Doe), who Kaimowitz claimed was being unlawfully held in a clinic for the purpose of experimental psychosurgery. The patient had been charged with the murder and rape of a nursing student at the state mental hospital where he had been confined. As a result, he was committed by a county court in 1955 to the state hospital as a criminal sexual psychopath, without a trial of the criminal charges, under the provisions of the law that existed at that time. In 1972, several physicians at the clinic where he was held filed a proposal to treat "uncontrollable aggression" with psychosurgery. The study was funded by the state legislature and had been approved by a Scientific Review Committee consisting of a professor of law and psychiatry, a representative from the clergy, and a certified pubic accountant. The patient's parents had given their permission for the experiment to proceed and the patient himself had signed an "informed consent" form, prior to his transfer to the clinic from the state mental hospital, in which he had agreed to be an experimental subject.

In reviewing the facts of the case, the court found that parental consent is ineffective in situations involving psychosurgery. The court noted that because there was a lack of knowledge with respect to the risks and results of psychosurgery, knowledgeable consent to such a procedure was impossible. Thirdly, the court observed that all of the patient's decisions had been made for him for 17 years by the hospital staff, without any opportunity to participate in the decision making process. The court characterized this environment as one which was "inherently coercive." Ultimately, the court concluded that the surgery could not proceed because the requisite elements of informed consent—competency, knowledge, and voluntariness—were absent (*Kaimowitz v. Department of Mental Health for the State of Michigan*, 1973).

IRBs have also been found to have little influence on the readability and understandability of informed consent forms by uneducated research participants (Hammerschmidt & Keane, 1992), and are generally unable to ensure that the researchers actually utilize only the forms and procedures that have been approved for use (Adkinson, Starklauf, and Blake, 1983; Delgado and Leskovac, 1986. See the discussion of intentional torts in chapter 5). The lack of standardization between IRBs (Castronovo, 1993) may create the appearance of injustice (Rosenthal and Blanck, 1993). Current procedures have also been criticized for their lack of a remedy to a researcher's violation of a protocol, apart from termination of funding by the funding source (Delgado and Leskovac, 1986). And, although IRBs appear to ensure that privacy is safeguarded, they are not always effective in assessing risks and benefits (Rosnow, Rotheram-Borus, Ceci, Blanck, and Koocher, 1993).

Conversely, other critics have charged that IRBs are often overzealous in acting as gatekeepers, at the expense of the scientists, who are ethically bound to do good research (Rosenthal and Rosnow, 1984). In practice, IRBs of medical schools are more likely to review the science more critically than are IRBs located in liberal arts colleges (Rosnow, Rotheram-Borus, Ceci, Blanck, and Koocher, 1993).

Several writers (Reiser and Knudson, 1993) have suggested the development of a position of "research intermediary" as a possible solution to some of the IRBs' systemic shortcomings. The research intermediary would assure that the research participants understand the research process by discussing with them the informed consent forms both before and after they have signed them. The intermediary would also monitor how well the research protocol was being followed. The intermediary would be hired and trained by the IRB, and would be charged with the responsibility of reporting directly to the IRB. Other suggestions made to improve IRB functioning have included the revision of federal requirements to allow IRBs greater flexibility and mandate increased accountability, the provision of training in research ethics to both investigators and IRB members, the establishment of more than one IRB at an institution, the reliance by IRBs on professional staff, the inclusion on an IRB of at least one voting member who acts as a representative of the participants and is independent of the research and the investigators (Moreno, 1998), the enhancement of the local IRB's ability to revise protocols in multisite studies, the inclusion on an IRB of persons knowledgeable about disabled persons needs where the IRB reviews a significant number of protocols involving persons with disabilities, and the mandated disclosure by investigators to IRBs of any actual or potential conflicts of interest in conducting the proposed research (Moreno, Caplan, Wolpe, and Members of the Project on Informed Consent, 1998).

One Institutional Review Board: An Example

This section presents the policies and procedures of one hospital-based IRB. The IRB is specifically charged in its statement of policies and procedures to
1. protect human subjects from undue risk and a deprivation of human rights,
2. insure that participation is voluntary,
3. maintain a balance between the risks and benefits of participation,
4. determine whether the research design and methods are appropriate to the objectives of the research,
5. protect the investigator by peer review and institutional approval to minimize risks of litigation, and
6. insure compliance with federally-mandated protocols.

The IRB meets twice a month to review protocols that were submitted at least two weeks prior to the regularly scheduled meeting. Members of the IRB include representatives from the hospital's clinical departments, the nursing department, the pharmacy, the affiliated medical school, the affiliated nursing school, a member from the community, and a member of the clergy. Members generally serve a three-year term, and may be re-appointed by the chairperson of the IRB. Members of the IRB who are involved in a project presented for review are explicitly prohibited from participating in the review of that project. Certain types of projects may receive expedited review. The IRB has chosen not to categorically exempt from review any studies involving human subjects. For each protocol reviewed, one of the following actions will be taken: approval, conditional approval pending receipt of statements pursuant to IRB recommendations, deferral pending resolution of questions raised by the IRB, or disapproval.

All protocols must be signed by the principal investigator, and carry the approval and signature of the department chairperson before being submitted to the IRB for review. Special reviews are conducted if the protocol involves experimental or investigational use of a radiographic study, radiation therapy, or radionuclides, if the protocol involves patient contact with new or non-standard electrical equipment, or if the protocol involves the use of investigational drugs. There are specific provisions that apply to questionnaire studies, blood drawing studies, and studies involving investigational drugs. Patients' charts may be accessed for the purpose of conducting research in compliance with the parameters instituted by the hospital's health information services department. If the study involving chart reviews will also require any form of contact with the patients or a member of the patient's family, the protocol must be submitted to the IRB for review and approval. The IRB's primary concern in such instances is the potential invasion of privacy and the use of confidential information.

In an emergency, when a research protocol requires immediate consideration for one patient prior to IRB review, the investigator must receive departmental approval and subsequent review by the IRB. The IRB must be notified within 5 working days. Pursuant to requirements of the Food and Drug Administration, the investigator must inform the IRB of the use of any experimental drug within 5 working days and must keep in a file a copy of the notification letter, the consent form, and the acknowledgement memo from the IRB.

Investigators are responsible for reporting immediately to the IRB any adverse events involving risks to human subjects or any unexpected results. The IRB is responsible for the reporting of serious adverse events to the FDA where the nature of the study is such that it falls within its jurisdiction. The IRB may revoke approval of a protocol if at any time the project is not in compliance with IRB guidelines.

The IRB has assumed a restrictive posture in the approval of research protocols involving normal children because of difficulties associated in obtaining truly informed assent from children. (See chapter 4 for additional discussion of consent and assent.) Studies that involve more than minimal risk and that will include legally incompetent persons as participants require written consent from the guardian or the person holding legal power of attorney for the participant. The signature of a spouse or relative is insufficient to authorize participation. Unconscious persons may not be used in research studies unless there is no alternative approved method or generally accepted therapy that provides an equal or greater likelihood of saving the patient's life.

Participants in the research may be paid for their participation, but not an amount so great that it could induce them to participate where they might not otherwise agree to do so. Participants must be advised if there will be any cost to them for their participation, such as for radiographic studies conducted solely for research purposes.

The consent form must contain statements relating to the purpose of the study, the approval of the patient's physician to contact the patient for participation, the procedures, the risks and benefits, alternatives to participation, the ability to withdraw from the study, any financial considerations, and the extent of confidentiality.

All protocols are reviewed at least annually. Failure to submit the required documentation for an annual review results in a notice of termination for lack of review. The IRB procedures explicitly advise investigators that the conduct of a study with active approval is illegal. Projects with high risks may be reviewed more frequently. The investigator is responsible for notifying the IRB upon discontinuation or completion of a project.

Institutional Review and International Research

Regulations governing research to be conducted outside of the United States by United States investigators require that the protocol be reviewed and approved not only by a review committee based in the United States, but also by the appropriate review committee in the country hosting the research. This requirement mirrors the provision found in the *International Guidelines for Ethical Review of Epidemiological Studies* (CIOMS, 1993: Guideline 48, at 23), which provides that the agency initiating the research submit the study protocol for ethical review; that the ethical standards utilized for that review be no less exacting than they would be for a study carried out in the initiating country; and that the ethical review committee in the host country assure itself that study meets own ethical requirements. A brief synopsis is provided below of the procedures used for ethical review in three other countries.

Australia

In Australia, ethical review committees are known as Institutional Ethics Committees (IECs). They operate under the National Health and Medical Research Council (NHMRC). The Council is charged with the responsibility of making recommendations to the government on matters relating to medical research. The Council has formulated specific guidelines for the conduct of medical research, guided in part by the Helsinki Declarations. The guidelines require that research participants be adequately informed of the risks and purposes of the research, that they consent to participation in writing, and that they be informed of their right to withdraw at any time. The National Bioethics Consultative Committee is responsible for advising the government on bioethical issues (Drahos, 1989).

Like the IRBs in the United States, the IECs have been criticized for their inability to sanction researchers who fail to comply with the approved protocol or to the established guidelines for research. The only real sanction is a financial one, through the withdrawal of research funding (Drahos, 1989).

The Netherlands

Research Ethics Committees (RECs) began to develop in the Netherlands in the 1970s and 1980s. A 1984 Order provided for the establishment of RECs in connection with hospitals, but specifically provided that several hospitals may come under one committee for the purpose of ethical review of research. The Order also requires that hospitals guarantee that a refusal by an individual to participate in a research study will not influence his or her entitlement to optimum care.

A study of RECs found that most experimental research is conducted in academic and general hospitals. A small proportion is carried out in specialized and psychiatric hospitals. Most institutions where the research is conducted have mechanisms for the scientific review of the proposals. Most of the research institutions (78%) carry out their own ethical review, often by an REC, but sometimes by medical staff or medical management.

Membership of an REC varies in size from 3 to 13; medical specialists are heavily represented on the RECs. One study found that about half of the RECs include a member of the clergy or an ethics specialist. Lawyers are most often members of academic RECs. The RECs rarely include nonprofessional representatives of consumer groups or patients' organizations (Bergkamp, 1988).

RECs have been criticized for their lack of legal authority, for the lack of clarity with respect to the scope and depth of their review, and for their inability to ensure that investigators are complying with the approved protocol (Bergkamp, 1988). One writer has suggested the formation of regional RECs to increase their abilty to review proposals independently and to increase uniformity in decision making. This would concomitantly reduce the likelihood that investigators would search for a more permissive REC (Bergkamp, 1988).

Uganda

Uganda's *Guidelines for the Conduct of Health Research Involving Human Subjects in Uganda* (*Guidelines*) establish multiple levels of review, beginning at the institutional level with institutional review committees (IRCs) and extending to the AIDS Commission for HIV-related research and to the National Council for Science and Technology (NCST) for all research, including that which is HIV-related. The regulations are intended to apply not only to universities and governmental entities, but also to nongovernmental organizations that may conduct health-related research, although not labeled as such, in conjunction with their program development or evaluation. Pursuant to the *Guidelines*, IRCs must now consist of at least 5 members of diverse tribes, religions, professions, and socioeconomic status. At least one member of the IRC may not be otherwise affiliated with the research-sponsoring institution in any way. Membership will rotate periodically and may be involuntarily terminated where it is established that an individual has committed misconduct or has refused to excuse him- or herself from discussion where there exists a conflict of interest.

IRC members are charged with the responsibility to ensure that any approved protocol (1) minimizes risks to the study participants, (2) adequately reflects and balances the risks and anticipated benefits to the participants, (3) provides for the equitable selection of research participants, (4) requires both the receipt and documentation of informed consent from each individual participant, (5) establishes adequate mechanisms to ensure the confidentiality of the data and the privacy of the participants, (6) establishes procedures to assure the safety of the participants, and (7) establishes additional safeguards for the protection of participants vulnerable to coercion or undue influence.

IRCs have authority to temporarily and provisionally suspend or terminate research that is in violation of the IRC's original approval or that has been associated with unexpected serious harm to the research participants. Each institution maintaining an IRC must provide written assurances to the NCST regarding the principles to be followed by the institution to protect the rights and welfare of the participants, the procedures to be followed by the IRC, and the composition of the IRC. This assurance is comparable to the Assurance of Compliance now required by the Office of Protection from Research Risks of the

U.S. Department of Health and Human Services of institutions conducting research with HHS funds, discussed in chapter 5.

Subsequent to local IRC review, a protocol must be submitted for review and receive the approval of the NCST. Although the scope of that review is similar to the review provided by the local IRC, review at a national level accomplishes several functions not attainable through local level review. Ethical review of the proposed research at a national level is intended to permit the consolidation of experts in a particular field to review the proposal and allow the NCST to approve those protocols that are consistent with and most likely to achieve the nation's health research objectives.

Exercise

You have been called in as a consultant to a public health department in a large urban area. A segment of the health department is concerned about the potential transmission of infectious diseases, such as HIV, hepatitis B and C, syphilis, and bacterial endocarditis through various segments of the population as the result of transmission between injecting drug users and their sexual and needle-sharing partners. Recent estimates indicate that there are approximately 20,000-30,000 individuals in the city who are regularly injecting illicit drugs. However, there are only 1,200 publicly funded treatment slots, and those slots are available to only individuals who are using heroin exclusively. Consequently, there are no publicly funded treatment slots for individuals injecting cocaine, methamphetamine, or other drugs.

In order to address these concerns, a health planner in the department of public health has designed a cross-sectional interview-based study of injection drug users' needle-sharing and sexual behaviors in order to better understand and estimate the potential rate of transmission throughout the population. Interviews are expected to last approximately one hour and will be conducted at the site at which a prospective participant is located, e.g. the street, a drug rehab center, a shelter, etc. As part of this study, the health department will also conduct street-based HIV testing of participants. Participants will be recruited through location-based sampling and snowball sampling. Each participant will be paid $50.00 for his or her participation in the study. Because the department is not university-based, it does not have an IRB.
1. What are the potential ethical implications of proceeding with this study in the absence of review by an IRB or similar body?
2. Assume for the purpose of this subpart only that you are a member of a university-based IRB, with which the public health department has contracted to review this proposed study. What, if any, concerns do you have regarding the study as it is currently formulated?

CONFLICTS OF INTEREST

A conflict of interest in the context of epidemiology has been defined as occurring

> whenever a personal interest or a role obligation of an investigator conflicts with an obligation to uphold another party's interest, thereby compromising normal expectations of reasonable objectivity impartiality in regard to the other party. Such circumstances are almost always to be scrupulously avoided in conducting epidemiologic investigations.
>
> Every epidemiologist has the potential for such conflict. An epidemiologist on the payroll of a corporation, a university, or a government does not encounter a conflict merely by the condition of employment, but a conflict exists whenever the epidemiologist's role obligation or personal interest in accommodating the institution, in job security, or in personal goals compromises obligations to others who have a right to expect objectivity and fairness. (Beauchamp et al., 1991)

Despite its reference to epidemiologists, the definition is equally valid for other scientific researchers. Although a conflict of interest may ultimately lead to scientific misconduct, the presence of one does not necessarily foresee the occurrence of the latter. (Scientific misconduct is discussed in chapter 5.)

Researcher conflict of interest may stem from any of several motivating forces, including altruism, a desire for personal recognition, or the possibility of financial reward. Conflicts of interest are of concern because they may introduce a bias into the research. Chalmers (1983) defined bias as "unconscious distortion in the selection of patients, collection of data, determination of final end points, and final analysis." The introduction of bias as the result of the investigator's conflict of interest can ultimately result in deficiencies in the design, data collection, analysis, or interpretation of a study. Investigator bias could potentially cause harm to the research participants and to those individuals who later rely on the research findings in making their own decisions about treatment or therapy. Even when there is no actual conflict of interest, a perceived conflict of interest may result in the erosion of trust (Beauchamp et al., 1991) of the public or the participants in the research, the research institution, or the investigators. These concerns are reflected in the *International Guidelines for Ethical Review of Epidemiological Studies* (CIOMS, 1991: guideline 31, at 18), which notes that "[h]onesty and impartiality are essential in designing and conducting studies, and presenting and interpreting findings...."

Financial Conflicts

Issues of financial conflict of interest have been the most visible form of conflict of interest in research and is, perhaps, the type of conflict that is most likely to lead to scientific misconduct (see chapter 5). The difficulty of maintaining one's objectivity in research while simultaneously maintaining an economic interest in the outcome of that research has been almost universally accepted. Relman (1989) commented:

> [i]t is difficult enough for the most conscientious researchers to be totally unbiased about their own work, but when an investigator has an economic interest in the outcome of the work, objectivity is even more difficult.

The Council on Scientific Affairs and the Council on Ethical and Judicial Affairs of the American Medical Association (1990) agreed:

> For the clinical investigator who has an economic interest in the outcome of his or her research, objectivity is especially difficult. Economic incentives may introduce subtle biases into the way research is conducted, analyzed, or reported, and these biases can escape detection by even careful peer review.

Wells (1987) has asserted that "physicians responsible for the care of the patients or subjects in [clinical studies] should not have a significant financial interest in the company or organization." The *International Guidelines for Ethical Review of Epidemiological Studies* (CIOMS, 1991) recognizes the potential for a conflict of interest even in the context of epidemiological studies and the difficulties that may result:

> Epidemiological studies may be initiated, or financially or otherwise supported, by governmental or other agencies that employ investigators. In the occupational and environmental health fields, several well-defined special interest groups may be in conflict: shareholders, management, labour, government regulatory agencies, public interest advocacy groups, and others. Epidemiological investigators may be employed by any of these groups. It can be difficult to avoid pressures resulting from such conflict of interest, and consequent distorted interpretations of study findings. Similar conflict may arise in studies of the effects of drugs and in testing medical devices. (Guideline 28, at 18)

Despite the widespread recognition that a financial interest may not be in the best interests of the science, researchers and their institutions are often tied to industry through financial considerations. A 1994 survey of life science companies revealed that 59 percent supported research based in academic institutions, although that funding constituted only 12 per cent of all university research in the life sciences (Blumenthal, 1995). Such relationships have been championed as increasing the pool of available funding and creating new opportunities. Critics have argued that the danger of such partnerships at the institutional level, rather than at the individual level, is

> the loss of the soul of the university as a reservoir of independent minds, who can freely and securely offer a critical perspective on the conditions and directions of society, including its technological, political, economic, and social organization. The university is composed of tenured prima donnas who speak their

mind and don't speak on behalf of their institution. This is a national resource for society. (Krimsky, 1995).

Researchers have not infrequently maintained significant financial interests in the products that they are testing or in the company that is funding their research. A recent examination of practices in the field of interventional cardiology found that "the system used to determine which device is best for heart patients can be influenced as much by personal financial interests as by scientific data" (Eichenwald and Kolata, 1999: A1). For instance, one researcher's dual role as a multimillion dollar investor in Heart Technology Inc. and as a researcher evaluating its product "Rotablator" prompted review by congressional leaders concerned about conflicts of interest in scientific research (Dalton, 1994). A survey of faculty receiving funding from the National Institutes of Health for their research found that almost 20 percent of the respondents had delayed the publication of their findings for six months or more in order to allow for patent application, to allow sufficient time to negotiate a patent, or to resolve dispute regarding ownership. Participation in an academic-industry research relationship and in the commercialization of university research were significantly associated with such delays (Blumenthal et al., 1997). A study of 800 articles appearing in leading scientific journals in 1992 found that 15.3 percent of the 1,105 lead authors had a financial interest in at least one of the articles, while lead authors of 34 percent of 789 papers had a financial interest in the research that they were describing (Wadman, 1997b). "Financial interest" in this study had even been defined somewhat restrictively:

> The author being listed as an inventor in a patent or patent application closely related to the published work . . . ; serving on a scientific advisory board of a company developing products related to the author's expertise...or serving as an officer or major shareholder of a company with commercial interests related to the research....Consultancies, personal financial holdings and honoraria were excluded, on the grounds that such links could not be adequately documented (Wadman, 1997b: 376)

Financial conflict of interest can take a number of forms, some of which are more subtle than others. For instance, those with a financial interest in the outcome of a particular study may be responsible for the funding of the study. This was the situation, for instance, in studies funded by tobacco companies to understand the role that the level of nicotine played in the maintenance of the smoking behaviors and the relationship between smoking and the development of lung cancer (Glantz et al., 1996) and in research funded by an asbestos manufacturer to assess the relationship between exposure to tremolite asbestos and the development of mesothelioma (Egilman, Wallace, and Horn, 1998). This is more likely to occur, however, in the context of clinical trials, where the clinical trial is funded by the pharmaceutical company developing and testing the new product. The consequences of such funding arrangements can be serious. For instance, companies sponsoring research have included clauses in the research contract giving them the right to veto publication of the results or to protect "trade secrets," and they have exercised that contractual right when it has been in their best financial interest to do so (Roush, 1997; Anon., 1996; Wadman, 1996). A number of leading journals maintain policies prohibiting editorial authors from having financial interests in any products discussed in the editorial, resulting, in effect, in a prohibition against the publication of the writing, even if the author discloses the conflict (Wadman, 1997a). Others may ban publication of any article where it appears that there may be a conflict of interest (Roberts and Smith, 1996).

Pharmaceutical companies may also offer investigators an opportunity for contract work. This may be beneficial to the investigator because he or she will not have to compete for funding through the peer-review process (Shimm and Spece, 1991b). The pharmaceutical company and the investigator may agree that the investigator will enter patients meeting specific entry criteria into a protocol written by the pharmaceutical company to satisfy the requirements of the Food and Drug Administration for a Phase I or Phase II trial. The investigator may receive a cash payment per patient enrollment, rather than a fixed amount that is unrelated to the number of patients enrolled. Investigators may also receive funding for travel to scientific meetings or consultantships (Shimm and Spece, 1991a).

Trial participants may be able to access drugs that would not otherwise be available to them. However, it is also possible that patients may be entered into a clinical trial when it is not in their best interest (Roizen, 1988), such as when an already-existing medication would be most effective for the treatment of their condition. This situation may be more likely to occur when the investigator is to receive a per-patient incentive and the capitation payments are small. This would require the investigator to enroll a greater number of patients to cover their fixed costs (Roizen, 1988). A study of physicians' and patients' attitudes towards physician enrollment of patients in postmarketing phase IV trials sponsored by pharmaceutical companies found that both physicians and patients believed that the physicians' judgment might be compromised with such a fee arrangement (La Puma, Stocking, Rhoades, and Darling, 1995).

A conflict may also arise in the context of royalty payments. For instance, a researcher may develop a new drug that may be used in the treatment of a specific disorder. The researcher then sells the product to a company in exchange for royalties. The company then asks the researcher to evaluate the new product in a clinical trial. The researcher's loyalties are divided between those to the research participants and those to the company that intends to market the product. One study has suggested that studies funded by pharmaceutical companies are more likely to favor the new therapy than studies funded through other sources (Davidson, 1986).

The ownership of stock in a company may constitute a conflict of interest. Lichter (1989) has asserted that "owning stock in a company at the same time one is conducting research, the results of which can affect the stock's value, creates the potential for bias, whether intended or not...." The potential for conflict of interest may not be as great where the investigator is examining a product for a very large company and the success or failure of that product is unlikely to have a large impact on the value of the company's stock. The success of even one product may determine the financial fate of a smaller company (Porter, 1992).

Margolis (1991) has argued from both a deontological and a utilitarian view that gifts in any form from a pharmaceutical company to a physician violate the fundamental duties of the physician with respect to nonmaleficence, fidelity, justice, and self-improvement. He maintains that the acceptance of a gift in any form, regardless of the value of that gift, transforms the physician into an agent of the company. A conflict of interest between the duty owed to the company is created, and that conflict is generally never disclosed to the patient. Similarly, Chren and colleagues have asserted that "[b]y offering a gift to another, a person is really proffering friendship, a relationship" (Chren, Landefeld, and Murray, 1989: 3449). Consequently, the acceptance of a gift signals

> the initiation or reinforcement of a relationship and it triggers an obligatory response from the recipient…the recipient generally assumes certain social duties such as grateful conduct, grateful use, and reciprocation. (Chren, Landefeld, and Murray, 1989: 3449)

These same arguments are relevant in the research context, where a researcher may have a conflict of interest between the duty owed to the company and the duty owed to the research participants.

Other Conflicts

Altruism

Although not often mentioned, a conflict of interest may arise from altruistic motives. For instance, a physician-researcher may be conducting a clinical trial to investigate a new drug to reduce the progression of a chronic disease. He is concerned for the health of his patient, who has tried all known and existing remedies, to no avail. Although he does not know whether or not the experimental drug is better than existing treatment, he overlooks the randomization procedure that has been established for the study (see appendix) and directs the individual to the group receiving the experimental drug, rather than the group receiving the conventional therapy. In this case, the researcher's desire to "do good" has overwhelmed his desire to seek the scientific truth (Porter, 1992). In some instances, this conduct may lead the investigator to draw erroneous conclusions from the data and may cause him or her to neglect or overlook important subissues.

Recognition

A researcher's desire for fame or professional recognition may also create a conflict of interest (Porter, 1992). For instance, a researcher's career may be enhanced greatly if he or she can claim to have found a cure for cancer or a vaccine for heretofore incurable and unpreventable diseases. Recognition may bring with it grant awards, financial remuneration, and the Nobel Prize.

A conflict of interest stemming from a desire for greater recognition may be manifested in more subtle ways. For instance, a reviewer of a manuscript submitted to a journal for publication may recommend against its publication not because of the quality of the manuscript or the researcher, but rather because the reviewer is also working in the same field and wishes to have his or her work recognized as the first to make a particular discovery. Similarly, a researcher serving as a reviewer of grant applications for a foundation or government funding entity may give a grant proposal a low score because he or she is planning on submitting, or has pending, a grant that addresses the same subject matter. The *Journal of the American Medical Association* has recognized this possibility:

> The Journal believes...that the term "conflict of interest" should apply not only to the possibility of financial gain for referees, but also to other, though less easily measurable, interests beyond the financial, such as the possibility of otherwise unmerited gains in priority of publication, personal recognition, career advancement, increased power, or enhanced prestige. (Southgate, 1987)

Strategies to Address Conflicts Before the Study Begins

Self-Imposed Restrictions

Self-elimination from participation in potentially conflicting activities may be critical to self-regulation. As an example, one research team investigating post coronary-artery-bypass graft (CABG) surgery treatments, developed the following restrictions:

> Investigators involved in the post CABG study will not buy, sell, or hold stock or stock options in any of the companies providing or distributing medication under study ...for the following periods: from the time the recruitment of patients of the trial begins until funding for the studying the investigator's unit ends and the results are made public; or from the time the recruitment of patients begins until the investigator's active and personal involvement in the study or the involvement of the institution conducting the study (or both) ends. (Healy et al., 1989: 951)

Similar restrictions were imposed on the acceptance of consultancy positions with the companies involved. Investigators were required to report on an annual basis participation in educational activities sponsored by the companies, participation in other research projects funded by the companies, and any uncompensated consulting provided by the researcher to the companies on unrelated issues.

Contract Negotiation

Contract negotiation may be an important mechanism to address potential conflicts of interest arising in the context of issues related to the ownership of the data and to sponsor control of the data and the research findings. In each of these instances, the research sponsor may choose to suppress the publication of the findings, believing that it is not in its best interest to allow them to be made public (Plant, Plant, and Vernon, 1996). The investigator, however, may have an interest in having the findings published, either because they represent a contribution to the field of interest, and/or for personal objectives, such as recognition within the field of endeavor. Wenger has speculated that there is an inherent tension between the cultures of researchers and funders that contributes to disagreements regarding the use of data and findings:

> On the one hand, at the level of basic value orientation, the academic ethos is one which places a high value on independence, intellectual autonomy, and creativity, while on the other, administrative ethos is one of agency loyalty, formal procedures and respect for authority. Related to the basic ethos are the hierarchical structures of the institutional backgrounds. The academic is part of a collegiate structure, loosely defined in terms of administrative responsibility rather than authority. As an academic, her or his professional responsibilities are comparable irrespective of her or his administrative responsibilities. Within the academic community, the researcher may have high or low status. The administrator, on the other hand, is part of a bureaucracy with (many) hierarchical strata. (Wenger, 1987: 211-212, citing Sharpe, 1978)

The investigator may attempt to negotiate with the potential funder for ownership rights to the data and the right to publish the findings without regard to the sponsor's agreement with the findings. In some cases, it may be possible to negotiate a right of review of manuscripts for the sponsor, without also giving the sponsor the right of veto. In some situations however, the investigator will have to decide whether or not to accept the funding despite the sponsor's refusal to provide funding without a right of veto. In such cases, "[a] good guide when you are facing a difficult decision is to consider whether you would be happy to be questioned about the decision on live television." (Smith, 1994: 1597)

Disclosure to Research Participants and Review Boards

Shimm and Spece (1991a) have argued, based on analogy to case law relating to patients and physician compensation, that research participants must be informed of all financial arrangements between the clinical researchers and manufacturers. The legal implications of a failure to inform the participants are addressed in chapter 5. Shimm and Spece (1991b) have also suggested that monies available from contracts with pharmaceutical companies for research that are in excess of the direct costs of a study be placed in an institution-wide pool. Investigators associated with the institution could compete on a local level for access to the pooled funds for their research.

Various international documents governing research are cognizant of the potential for a conflict of interest and are forceful in their injunction to disclose all such relationships. The *International Guidelines for Ethical Review of Epidemiological Studies* (CIOMS, 1991: 18-19) provides in Guideline 27

> It is an ethical rule that investigators should have no undisclosed

conflict of interest with their study collaborators, sponsors, or subjects. Investigators should disclose to the ethical review committee any potential conflict of interest. Conflict can arise when a commercial or other sponsor may wish to use study results to promote a product or service, or when it may not be politically convenient to disclose findings.

Peer Review

Peer review of the ethical issues involved in a research protocol often occurs when the protocol is submitted to a funding source for review. For instance, a review committee of the National Institute of Health would review not only the scientific aspects of a research protocol, but also the ethical aspects, such as the mechanism for obtaining informed consent and the strategies developed to ensure confidentiality of the data. The *Guidelines for Good Clinical Practice for Trials on Pharmaceutical Products* (WHO, 1995, section 4.11, at 115) recognize the critical role of review committees and requires that "[t]he relationship between the investigator and the sponsor in matters such as financial support, fees and honorarium payments in kind [be] stated in writing in the protocol or contract." The *International Guidelines for Ethical Review of Epidemiological Studies* (CIOMS, 1991: 18-19) provide in Guideline 29:

> Investigators and ethical review committees will be sensitive to the risk of conflict and committees will not normally approve proposals in which conflict of interest is inherent. If, exceptionally, such a proposal is approved, the conflict of interest should be disclosed to prospective subjects and their communities.

Peer review has been characterized as a "key form of control" (Shipp, 1992). It is helpful because the work is reviewed by others who presumably have greater skill and expertise in the same area of endeavor. However, peer review cannot be the sole mechanism for detection of and attention to a conflict of interest. Many conflicts of interest may not be visible to those reviewing the work (Council on Scientific Affairs and Council on Ethical and Judicial Affairs, 1990; Shipp, 1992).

Federal Requirements

The Public Health Service requires that recipients of its grants maintain conflict of interest policies. The 1990 PHS Grants Policy Statement requires that recipient organizations

> establish safeguards to prevent employees, consultants, or members of governing bodies from using their positions for purposes that are or give the appearance of being, motivated by a desire for private financial gain for themselves or others such as

those with whom they have family, business, or other ties. Therefore, each institution receiving financial support must have written policy guidelines on conflict of interest and the avoidance thereof. These guidelines should reflect State and local laws and must cover financial interest, gifts, gratuities and favors, nepotism, and other areas such as political participation and bribery. (Public Health Service, United States Department of Health and Human Services, 1990)

Institutional Controls and Review

Various models have been developed for the identification and review of conflicts of interest. Institutions may prohibit specific activities, classify activities by level of scrutiny and control, or review activities on an individual basis to determine whether a conflict exists (Shipp, 1992). Various models also exist for the disclosure of possible conflicts of interest, including annual submissions of disclosures or ad hoc submissions. A committee may be in place to review the disclosures and the potential for conflict of interest.

Institutions differ in both the prescribed manner of disclosure and what must be disclosed. The requirement of disclosure may be institution-initiated or investigator-initiated. The institution may require that any, or all, of the following be disclosed: outside professional positions, equity holdings, outside professional income, gifts, honoraria, or loans (Shipp, 1992).

Activities may be found to be unacceptable and therefore prohibited. For instance, some institutions have prohibited investigators from receiving funds from specified sources, due to a perceived conflict of interest between the goals of the research and the goals of the prospective research sponsors (Cohen, 1996).Some activities may be permissible, but may be problematic. The resolution of such situations could include the public disclosure of relevant information, the reformulation of the research, closer monitoring of the research, divestiture by the investigator of his or her personal interests, a cessation of the investigator's participation with the research project, the reduction of the investigator's involvement with the research project, the termination of inappropriate student involvement in the project, or the termination of outside relationships that introduce the conflict (Association of American Medical Colleges, 1990).

CHAPTER SUMMARY

This chapter has reviewed the ethical issues that may arise in the process of designing and initiating a study. These include, but are not limited to, the design of the study, the formulation of inclusion and exclusion criteria, the selection of the research team, the choice of a comparison group or treatment, and the balancing of the risks and benefits. Mechanisms for the minimization of ethical problems later on during the course of the study were discussed, including the review and approval of the protocol by an institutional committee and the elimination or modification of conflicts of interest that may prove to be problematic.

EXERCISE

Assume that you are an independent epidemiologist, working as a consultant on contracts of your choosing. At one time, you conducted studies for the U.S. military examining the effects of certain types of "biological warfare" on U.S. military personnel. You have since continued your work as a consultant for the U.S. military, for which you are paid a substantial fee. A group of military personnel has now raised with Congress the issue of their unknowing and unwilling exposure to certain types of "biological warfare," which were not the subject of your prior research. Congress has now called upon you to testify regarding the military's conduct of such experiments and the effects of the exposures on its members. For the purpose of this question, disregard any legal obligations that you may have in this situation.

1. What ethical issues are raised in this situation, apart from conflict(s) of interest?
2. What conflicts of interest exist in this situation? Are they role-related or value-related?
3. What ethical theories, principles, and rules must you consider in resolving the conflicts that you have identified?
4. Explain how you will apply the theories, principles, and rules in (3) above to resolve the conflicts of interest.

References

Adkinson, N.F., Starklauf, B.L., & Blake, D.A. (1983). How can an IRB avoid the use of obsolete consent forms? *IRB, 5,* 10-11.

Angell, M. (1997). The ethics of clinical research in the Third World. *New England Journal of Medicine, 337,* 847-849.

Anon. (1996, September 5). Tobacco and researchers' rights. *Nature, 383,* 1.

Association of American Medical Colleges. (1990). *Guidelines for Dealing with Faculty Conflicts of Commitment and Conflicts of Interest in Research.* Washington, D.C. AAMD Ad Hoc Committee on Misconduct and Conflict of Interest in Research.

Baron, R.A. (1981). The "costs of deception" revisited: An openly optimistic rejoinder. *IRB: A Review of Human Subjects Research 3,* 8-10.

Baumrind, D. (1979). IRBs and social science research: The costs of deception. *IRB: A Review of Human Subjects Research, 1,* 1-4.

Baumrind, D. (1978). Nature and definition of informed consent in research involving deception. In National Commission for the Protection of Human Subjects of Biomedical and Behavioral Research. *The Belmont Report: Ethical Principles and Guidelines for the Protection of Human Subjects of Research* (pp. 23.1-23.71). (DHEW Publication No. (OS) 78-0010). Washington, D.C.: Department of Health, Education, and Welfare.

Bayer, R. (1998). The debate over maternal-fetal HIV transmission prevention trials in Africa, Asia and the Caribbean: Racist exploitation or exploitation of racism. *American Journal of Public Health, 88,* 567-570.

Beauchamp, T.L. (1996). Moral foundations. In S.S. Coughlin & T.L. Beauchamp (Eds.). *Ethics and Epidemiology* (pp. 24-52). New York: Oxford University Press.

Beauchamp, T.L., Cook, R.R., Fayerweather, W.E., Raabe, G.K., Thar, W.E., Cowles, S.R., & Spivey, G.H. (1991). Ethical guidelines for epidemiologists. *Journal of Clinical Epidemiology, 44,* 151S-169S.

Bergkamp, L. (1988). Research ethics committees and the regulation of medical experimentation with human beings in the Netherlands. *Medicine & Law, 7,* 65-72.

Berkowitz, L. (1978). Some complexities and uncertainties regarding the ethicality of deception in research with human subjects. In National Commission for the Protection of Human Subjects of

Biomedical and Behavioral Research. *The Belmont Report: Ethical Principles and Guidelines for the Protection of Human Subjects of Research* (pp. 24.1-24.34). (DHEW Publication No. (OS) 78-0010). Washington, D.C.: Department of Health, Education, and Welfare.

Blumenthal, D. (1995). Academic-industry relationships in the 1990s: Continuity and change. Oral presentation at symposium, Ethical issues in research relationships between industry, Baltimore, Maryland. Cited in J.T. Rule & A.E. Shamoo. (1997). Ethical issues in research relationships between universities and industry. *Accountability in Research, 5*, 239-249.

Blumenthal, D., Campbell, E.G., Anderson, M.S., Causino, N., and Louis, K.S. (1997). Withholding research results in academic life science: Evidence from a national survey of faculty. *Journal of the American Medical Association, 277*, 1224-1228.

Castronovo, F.P., Jr. (1993). An attempt to standardize the radiodiagnostic risk statement in an institutional review board consent form. *Investigative Radiology, 28*, 533-538.

Chalmers, T.C. (1983). The control of bias in clinical trials. In *Clinical Trials: Issues and Approaches* (pp. 15-127). New York: Dekker.

Chren, N.M., Landefeld, C.S. & Murray, T.H. (1989). Doctors, drug companies, and gifts. *Journal of the American Medical Association, 262*, 3448-3481.

Classen, W.H. (1986). Institutional review boards: Have they achieved their goal? *Medicine & Law, 5*, 387-393.

Cohen, J. (1997, May 16). Ethics of AZT studies in poorer countries attacked. *Science, 276*, 1022.

Cohen, J. (1996, April 26). Tobacco money lights up a debate. *Science, 272*, 488-494.

Coughlin, S.S. (1996). Ethically optimized study designs in epidemiology. In S.S. Coughlin & T.L. Beauchamp, (Eds.). *Ethics and Epidemiology* (pp. 145-155). New York: Oxford University Press.

Council for International Organizations of Medical Sciences (CIOMS), World Health Organization (WHO). (1993). *International Guidelines for Biomedical Research Involving Human Subjects*. Geneva: WHO.

Council for International Organizations of Medical Sciences (CIOMS). (1991). *International Guidelines for Ethical Review of Epidemiological Studies*. Geneva: Author.

Dalton, R. (1994, April 4). 2 at UCSD banned from experimentation on patients. *San Diego Union Tribune*, B-1.

Davidson, R.A. (1986). Sources of funding and outcome of clinical trials. *Journal of General Internal Medicine, 3*, 155-158.

Delgado, R. & Leskovac, H. (1986). Informed consent in human experimentation: Bridging the gap between ethical thought and current practice. *UCLA Law Review, 34*, 67-130.

Drahos, P. (1989). Ethics committees and medical research: The Australian experience. *Medicine & Law, 8*, 1-9.

Egilman, D., Wallace, W., & Horn, C. (1998). Corporate corruption of medical literature: Asbestos studies by W.R. Grace & Co. *Accountability in Research, 6*, 127-147.

Eichenwald, K. & Kolata, G. (1999, November 30). When physicians double as businessmen. *New York Times*, A1.

Food and Drug Administration (FDA). (1998, September). Information Sheets: Guidance for Institutional Review Boards and Clinical Investigators, 1998 Update. Available at http://www.fda.gov/oc/oha/IRB/toc10.html.

Gambia Government/Medical Research Council Joint Ethical Committee. (1998). Ethical issues facing medical research in developing countries. *Lancet, 351*, 286-287.

Glantz, S.A., Slade, J., Bero, L.A., Hanauer, P., & Barnes, D.E. (1996). *The Cigarette Papers*. Berkeley, California: University of California Press.

Goffman, E. (1961). *Asylums: Essays on the Social Situation of Mental Patients and Other Inmates*. New York: Doubleday.

Hammerschmidt, D.E. & Keane, M.A. (1992). Institutional review board (IRB) review lacks impact on the readability of consent forms for research. *American Journal of the Medical Sciences, 304*, 348-6351.

Healy, B.M., Campeau, L., Gray, R., Herd, A., Hoogwerf, B., Hunninghake, G., Stewart, W., White, C., & the Investigators f the Post Coronary Artery Bypass Graft Surgery Clinical Trial. (1989). Special report: Conflict-of-interest guidelines for a multicenter clinical trial of treatment after coronary-artery bypass-graft surgery. *New England Journal of Medicine, 320*, 949-951.

Institute of Medicine. (1994). *Women and Health Research*. Vol. 1. *Ethical and Legal Issues of Including Women in Clinical Studies*. Washington, D.C.: National Academy Press.

Kaimowitz v. Department of Mental Health for the State of Michigan, No. 73-19434-AW (Mich. Cir. Ct. Wayne County, July 10, 1973).

Kaufman, R.E. (1994). Scientific issues in biomedical research with children. In M.E. Grodin & L.H. Glantz (Eds.). *Children as Research Subjects: Science, Ethics & Law.* (pp. 29-45). New York: Oxford University Press.

Krimsky, S. (1995). An evaluation of disclosure of financial interests in scientific publications. Oral presentation at symposium, Ethical issues in research relationships between universities and industry, Baltimore, Maryland. Cited in J.T. Rule & A.E. Shamoo. (1997). Ethical issues in research relationships between universities and industry. *Accountability in Research, 5,* 239-249.

La Puma, J., Stocking, C.B., Rhoades, W.D., & Darling, C.M. (1995). Financial ties as part of informed consent to postmarketing research: Attitudes of American doctors and patients. *British Medical Journal, 310,* 1660-1661.

Levine, R.J. (1996). International codes and guidelines for research ethics: A critical appraisal. In H.Y. Vanderpool, (ed.). *The Ethics of Research Involving Human Subjects: Facing the 21st Century* (pp. 235-259). Frederick, Maryland: University Publishing Group.

Levine, R.J. & Cohen, E.D. (1974). The Hawthorne effect. *Clinical Research, 22,* 111-112.

Lichter, P.R. (1989). Biomedical research, COI, and the public trust. *Ophthalmology, 96,* 575-578..

Lo, B. (1990). Assessing decision-making capacity. *Law, Medicine & Health Care, 18,* 193-201.

Loue, S. (in press). Elder abuse in medicine and law: The need for reform. *Journal of Legal Medicine.*

Loue, S., Okello, D., & Kawuma, M. (1996). Research bioethics in the Uganda context: A program summary. *Journal of Law, Medicine, & Ethics, 24,* 47-53.

Lurie, P. & Wolfe, S.M. (1997). Unethical trials of interventions to reduce perinatal transmission of human immunodeficiency virus in developing countries. *New England Journal of Medicine, 337,* 853-856.

Macklin, R. (1989). The paradoxical case of payment as benefit to subjects. *IRB, 11,* 1-3.

Margolis, L.H. (1991). The ethics of accepting gifts from pharmaceutical companies. *Pediatrics, 88,* 1233-1237.

Mead, M. (1970). Research with human beings: A model derived from anthropological field practice. In P.A. Freund (Ed.). *Experimentation with Human Subjects* (p. 152-177). New York: George Braziller.

Moreno, J.D. (1998). IRBs under the microscope. *Kennedy Institute of Ethics Journal, 8,* 329-337.

Moreno, J., Caplan, A.L., Wolpe, P.R., & Members of the Project on Informed Consent, Human Research Ethics Group. (1998). Updating protections for human subjects involved in research. *Journal of the American Medical Association, 280,* 1951-1958.

Plant, M., Plant, M., & Vernon, B. (1996). Ethics, funding, and alcohol research. *Alcohol & Alcoholism, 31,* 17-25.

Porter, R.J. (1992). Conflict of interest in research: Personal gain—The seeds of conflict. In R.J. Porter & T.E. Malone, (Eds.). *Biomedical Research, Collaboration, and Conflict of Interest* (pp. 135-149). Baltimore: Johns Hopkins University Press.

Public Health Service, United States Department of Health and Human Services. (1990). Grants Policy Statement.

Reiser, S.J. & Knudson, P. (1993). Protecting research subjects after consent: The case for the "research intermediary." *IRB, 15,* 10-11.

Roberts, J. & Smith, R. (1996, January 20). Publishing research supported by the tobacco industry. *British Medical Journal, 312,* 133.

Roizen, B. (1988). Why I oppose drug company payment of physician/investigators on a per patient/subject basis. *IRB, 10,* 9-10.

Rosenthal, R. & Blanck, P.D. (1993). Science and ethics in conducting, analyzing, and reporting social science research: Implications for social scientists, judges, and lawyers. *Indiana Law Journal, 68,* 1209-1228.

Rosenthal, R. & Rosnow, R.L. (1984). Applying Hamlet's question to the ethical conduct of research: A conceptual addendum. *American Psychologist, 39,* 561-563.

Roush, W. (1997, April 25). Secrecy dispute pits Brown researcher against company. *Science, 276,* 523-524.

Rosnow, R.L., Rotheram-Borus, M.J., Ceci, S.J., Blanck, P.D., & Koocher, J.P. (1993). The institutional review board as a mirror of scientific and ethical standards. *American Psychologist, 48,* 821-826.

Sharpe, L.J. (1978). Government as clients for social science research. In M. Bulmer, ed. *Social Policy Research* (pp. 67-82). London: Macmillan.

Shimm, D.S. & Spece, R.G., Jr. (1991a). Conflict of interest and informed consent in industry-sponsored clinical trials. *Journal of Legal Medicine, 12,* 477-513.

Shimm, D.S. & Spece, R.G., Jr. (1991b). Industry reimbursement for entering patients into clinical trials: Legal and ethical issues. *Annals of Internal Medicine, 115,* 148-151.

Smith, T., Moore, E.J.H., & Tunstall-Pedoe, H. (1997). Review by a local medical research ethics committee of the conduct of approved research projects, by examination of patients' case notes, consent forms, and research records and by interview. *British Medical Journal, 314*, 1588-1590.

Shipp, A.C. (1992).How to control conflict of interest. In R.J. Porter & T.E, Malone (Eds.). *Biomedical Research, Collaboration, and Conflict of Interest* (pp. 163-184). Baltimore: Johns Hopkins University Press.

Smith, R. (1994). Editorial. Questioning academic integrity: Be prepared to defend yourself on television. *British Medical Journal, 309*, 1597.

Varmus, H. & Satcher, P. (1997). Ethical complexities of conducting research in developing countries. *New England Journal of Medicine, 337*, 1003-1005.

Wadman, M. (1996, May 2). Drug company 'suppressed' publication of research. *Nature, 381*, 4.

Wadman, M. (1997a, April 17). Journals joust over policy on authors' interests. *Nature, 386*, 634.

Wadman, M. (1997b). Study discloses financial interests behind papers. *Nature, 385*, 376.

Wells, F.O. (1994). Management of research misconduct—in practice. *Journal of Internal Medicine, 235*, 115-121.

Wenger, G.C. (Ed.). (1987). *The Research Relationship*. London: Allen and Unwin.

World Health Organization. (1995). *Guidelines for Good Clinical Practice (GCP) for Trials on Pharmaceutical Products*. WHO Technical Report Series, No. 850 (pp. 97-137).

45 Code of Federal Regulations §§ 46.201-.211, .301-.306, .401-.409 (1999).

4
ETHICAL ISSUES DURING AND AFTER THE STUDY

THE INFORMED CONSENT PROCESS

As we have seen, the Nuremberg Code specifies that participation of an individual in research requires voluntary consent:

> The voluntary consent of the human subject is absolutely essential.
> This means that the person involved should have legal capacity to give consent; should be so situated as to be able to exercise free power of choice, without the intervention of any element of force, fraud, deceit, duress, over-reaching, or other ulterior form of constraint or coercion; and should have sufficient knowledge and comprehension of the elements of the subject matter involved as to enable him to make an understanding and enlightened decision. The latter element requires that before the acceptance of an affirmative decision by the experimental subject there should be made known to him the nature, duration, and purpose of the experiment; the method and means by which it is to be conducted; all inconveniences and hazards reasonably to be expected; and the effects upon his health or person which may possibly come from his participation in the experiment.

The first Guideline of the *International Guidelines for Biomedical Research Involving Human Subjects* (CIOMS and WHO, 1993: 13) provides that:

> For all biomedical research involving human subjects, the investigator must obtain the informed consent of the prospective subject or, in the case of an individual who is not capable of giving informed consent, the proxy consent of a properly authorized representative.

And, as we have seen, the International Covenant on Civil and Political Rights, which is legally binding on signatory nations, has been interpreted as requiring informed consent.

The requirement of informed consent, however, is not limited to biomedical research. For instance, it also applies to research relating to peoples' attitudes towards disease and peoples' sexual behavior, both of which can be classified as health research, and neither of which would be thought of as biomedical research. Similarly, the *International Guidelines for the Ethical Review of Epidemiological Studies* (CIOMS, 1991) provides in its first Guideline:

> When individuals are to be subjects of epidemiological studies, their informed consent will usually be sought. For epidemiological studies that use personally identifiable private data, the rules for informed consent vary...Consent is informed when it is given by a person who understands the purpose and nature of the study, what participation in the study requires the person to do and to risk, and what benefits are intended to result from the study.

Why is informed consent so integral to each of these documents? We can look to each of the ethical theories for guidance. Deontology has as one of its major premises respect of the individual, i.e. not treating an individual merely as a means. Informed consent becomes a way of operationalizing that tenet, through an acknowledgement of an individual's values and choices that are freely made. Utilitarianism seeks to maximize good. Involvement of individuals in research without their understanding or their permission, or against their will, may lead to a distrust of science and scientists, harm to the individuals, and the questionable validity of the research findings—a result that could hardly be thought of as the maximization of good, by any definition. A casuistic analysis of past instances of experimentation involving human beings reveals considerable long-term and short-term harm to both individuals and the larger society in those instances in which research was conducted on unknowing and/or involuntary subjects. Consider, for instance, the Nazi experiments, the Tuskegee experiment, Willowbrook, and the Cold War radiation experiments, to name but a few. The principle that seems to emerge from such an analysis is congruent with that of deontology: respect for persons, which can be operationalized to informed consent. Communitarianism emphasizes communal values and relationships. Accordingly, it can be argued that the integrity of a community cannot be established and preserved, and its values enhanced, absent recognition of the integrity of its component parts. Virtue ethics emphasizes the qualities of respectfulness, nonmalevolence, and benevolence, which again argue for the recognition of and respect for an individual's freely made choice. This, again, can be operationalized as informed consent.

Integral to the concept of informed consent, as it has traditionally been applied by Western nations, is the concept of the individual as an autonomous being. This conceptualization of the individual may, however, be discordant "with more relational definitions of the person found in other societies . . . which stress the embeddedness of the individual within society and define a person by his or her relations to others" (Christakis, 1988: 34). In some cultures, decisions may be made in consultation with community leaders. Individuals may not even consider refusing a request to participate once permission has been given by a community leader or a family representative (Hall, 1989). Indeed, insistence on informed consent in a "westernized manner" without consideration of the cultural context may produce harm:

> [S]eeking informed consent to research from individuals [in certain developing world settings] may tend o weaken the social fabric of a nonindividualistic society, forcing it to deal with values it does not hold and possibly sowing disorder that the community will have to reap long after the investigators have gone home.... It is questionable that [our Western individualism] had been an unmitigated good for our own civilization and very questionable that it is up to standard for export. We ought, in truth, to be suitably humble about the worth of procedures [individual consent] developed only to cater to a very Western weakness....How can it be a sign of respect for people, or of our concern for their welfare, that we are willing to suppress research that is conducted according to the laws and cultures of the countries in which it is being carried out? (Newton, 1990:11)

Similarly, though, the position of ethical relativists—that actions are defined as right or wrong in the context of a specific culture and, consequently, are not subject to to judgment—may result in harm. After all, the Nazi experiments were conducted in accordance with laws that permitted such activity.

It is critical that consent procedures be adapted, as appropriate, to accommodate such differences, without abandoning the need for informed consent (WHO, 1989). Individual informed consent serves various purposes, including (1) serving as a reminder of the Kantian Categorical Imperative against using human beings merely as means; (2) requiring additional thoughtfulness on the part of the investigators by requiring them to explain the study; (3) regularizing relations between research participants and investigators; and (4) safeguarding individuals from invasions of their privacy (Capron, 1991).

As an example, Hall (1989) has reported on a model of informed consent used in Gambia, which appears to accommodate local concerns while still fulfilling the purposes of informed consent. Consent from the individual follows a series of permissions and negotiation of consensus, beginning with the government and then proceeding to the chief of the district, the head of the village, village meetings and, ultimately, to each individual to ask consent for participation. This resolution exemplifies the concept of ethical pluralism, which differs from both ethical universalism and relativism in four important respects: (1) it requires an ongoing dialogue between ethical systems; (2) it requires the negotiation between ethical ystems with regard to a specific situation; (3) it requires an assessment by each ethical system of itself and the "dissonant" system, and (4) it demands the acknowledgement that some ethical conflicts are irresolvable but must be addressed and dealt with nevertheless (Christakis, 1996). The *International Guidelines for Ethical Review of Epidemiological Studies* (CIOMS, 1991: Guideline 5, at 12-13) recognizes the need for such accommodations:

> When it is not possible to request informed consent from every individual to be studied, the agreement of a representative of a community or group may be sought, but the representative should be chosen according to the nature, traditions, and political philosophy of the community or group. Approval given by a

community representative should be consistent with general ethical principles. When investigators working with communities, they will consider communal rights and protection. For communities in which collective decision-making is customary, communal leaders can express the collective will. However, the refusal of individuals to participate in a study has to be respected: a leader may express agreement on behalf of a community, but an individual's refusal of personal participation is binding.

The various international documents enumerate the elements of informed consent: voluntariness, information, understanding, and the capacity to consent. Although each of these elements is discussed separately below, it is important to recognize that they are interwoven. Although recruitment relates to voluntariness, it is clearly related to the information provided. And, it is related to the principle of justice, as well as the principle of respect for persons. Similarly, the manner in which information is provided is relevant to the ability of the participants to understand. A well-designed informed consent process considers and incorporates the complexities and subtleties of such relationships in the context of the particular study's goals and procedures, while recognizing the concerns and characteristics of the participant and target populations. Suggestions for the development of an informed consent process are enumerated in Table 3.

Recruitment

Recruitment is intimately tied to issues relating to voluntariness, information, and justice. Consider, for instance, the studies that were reviewed in chapter 1, dealing with human experimentation. The subjects of Tuskegee were recruited into the study based on the belief, derived from the information that the investigator provided to them, that they would be receiving treatment for their disease. The parents of the children who underwent the hepatitis experiments at Willowbrook agreed to their children's participation based on incomplete knowledge of the risks and the benefits, as well as a promise that their child's admission to the hospital would be expedited if they gave their consent.

It is clear that barriers to recruiting may exist and may make it extraordinarily difficult to move forward with a study. These barriers are discussed below. Yet, in devising strategies to overcome these barriers, it is critical that concern for the participants be given priority over the conduct of the study.

Table 3. Considerations in the Development of an Informed Consent Process

Element	Items to be included:
Information	The study involves research.
	The purpose of the study
	The expected duration of the individual's participation
	A description of the procedures
	The identification of procedures that are experimental
	A description of any foreseeable risks or discomforts to the participant
	A description of any benefits to the participant that can be reasonably be expected from the research
	A statement disclosing appropriate alternative treatments or procedures that may be beneficial to the participant
	A description of the extent to which confidentiality will be maintained
	An explanation as to whether any compensation or medical treatment is available where the experiment involves more than minimal risk and where additional information may be obtained regarding the compensation or if injury occurs
	A statement that participation is voluntary
	A statement that the participant may withdraw at any time without penalty or loss of benefits to which he or she is otherwise entitled
	A statement that refusal to participate will not result in a loss of benefits to which the individual is otherwise entitled
	A statement of the circumstances under which participation may be terminated involuntarily (optional under federal regulations)
	The consequences of the participant's withdrawal and the procedures for such (optional under federal regulations)
	A statement that the procedure or treatment may involve risks to the participant, embryo, or fetus which are unforeseeable (as appropriate to study)
	Costs to the participant of withdrawal (as appropriate)
	A statement that significant new findings that develop during the course of the research and that may affect the participants' willingness to continue will be provided (optional under federal regulations)
Understanding	The written consent form or information sheet to be provided to the participant if consent is given orally is written in the language best understood by the participant
	The written consent form or information sheet is written at a reading level that can be understood by the participants.
	Alternative mechanisms are employed to convey the necessary information, as appropriate, e.g., video, an incremental

	information process.
Capacity	The participant has legal capacity to consent
	If the participant does not have legal capacity to consent, procedures have been implemented to obtain permission from the appropriate, authorized individual or entity and the participant has provided assent to the extent possible. (See text for discussion of when assent is required.)
	If the participant is cognitively impaired, he or she has been assessed for the extent of that impairment to determine if consent is possible.
	If the participant is cognitively impaired so that consent is not possible, permission to proceed has been obtained from an authorized individual or entity and additional protections have been implemented to protect the participant from harm.

Barriers to Recruitment

Access to Treatment. We discussed in chapter 3 various design issues that now become relevant to the issue of recruitment. As an example, suppose that you are conducting a Phase III clinical trial of a new drug, to be used to reduce the symptoms of a chronic disease. (See the appendix for a review of clinical trials.) A clinical trial means that individuals will be randomized to receive either the standard treatment or the experimental drug or, depending on the study design, will be randomized to receive either a placebo or the experimental drug. However, some individuals may be unwilling to leave their treatment and, consequently, their health outcome, to chance by being randomized to the experimental treatment or to the comparison/placebo group (Welton, Vickers, Cooper, Meade, and Marteau, 1999). For instance, individuals may be fearful of the toxic effects of the drug that is under investigation (Boffey, 1987). Recruitment for the initial studies of AZT therapy for HIV was slowed because of such concerns (Kolata, 1988). Recruitment may be hampered, as well, if the drug to be studied is widely available outside of the research setting. For instance, investigators had a great deal of difficulty recruiting participants for the initial trials of ddI because of the widespread availability of the drug outside of the study (Cimons, 1989).

Distrust of Researchers. Chapter 1 discussed in great detail some of the studies that have been conducted in the United States with apparent disregard of ethical standards and the welfare of the research participants. The conduct of research in this manner has left a lasting impression on many communities, resulting in a distrust of health researchers and an unwillingness to participate in research. The Tuskegee study, for instance, provides clear evidence to many individuals of the government's indifference to persons with black skin. One survey of 220 Afrcian Americans found that 43 percent believed that research in the United States is ethical, 11 percent believed that it was unethical, and 46 percent wanted more information but indicated that they were somewhat wary of research (Million-Underwood, Sanders, and Davis, 1993). A recent interview study of 1,882 patients in 5 geographic areas found that being African American was associated with the belief that medical research usually or always involves unreasonable risk and that patients usually or always are pressured into participating in research (Sugarman, Kass, Goodman, Parentesis, Fernandes, and Faden, 1998). Not surprisingly, some communities view AIDS as yet another genocidal effort directed at blacks (Cantwell, 1993; Guinan, 1993; Thomas and Quinn, 1991). A recent survey of 520 African Americans found that 27 percent agreed or somewhat agreed with the statement, "HIV/AIDS is a man-made virus that the federal government made to kill and wipe out black people" (Klonoff and Landrine, 1999). An additional 23 percent neither agreed nor disagreed. Those who believed that AIDS is a genocidal conspiracy were more likely to be male college graduates who had experienced frequent racial discrimination (Klonoff and Landrine, 1999).

Other communities have also been distrustful of government efforts to conduct research. Injecting drug users, for instance, may believe that the government continues to prohibit needle exchange programs, despite mounting evidence in support of their usefulness in reducing HIV transmission, and to underfund drug rehabilitation programs as part of a systematic effort to eliminate injecting drug users through their resulting deaths. Gay men were suspicious of government efforts to combat HIV as a result of the initial characterization of the disease as affecting only gay men, the labeling of the diseases as Gay Related Immune Disorder (GRID), and the perceived unwillingness of the federal government to fund adequately HIV research and prevention efforts (Cantwell, 1993).

Distrust of public health authorities has also resulted from what has been perceived as an attempt to exclude particular groups, such as women, injecting drug users, and persons of color, from participating in research that appears to be promising (Levine, 1989; Steinbrook, 1989). Non-English speakers have often been excluded because of the logistical difficulties and increased costs associated with the translation of forms, the interpretation of interviews, and the assurance of participant comprehension. Females have been systematically excluded because of reproductive considerations and the fear of liability should injury to the woman, fetus, or potential fetus occur (Merton, 1993).

Stigmatization. Recruitment for some studies may be hindered because of fears about the consequences of participating in a study. For instance, in some cultures, a fatal disease may be associated with witchcraft and sorcery, resulting in ostracism of the individual from his or her family and/or community (Schoepf, 1991). Women may be particularly vulnerable to adverse consequences. Responsibility for the disease and its transmission, such as HIV, may be attributed to the woman even though she may have contracted the infection from her male partner. Her participation in a study may result in rejection by her partner.

Economic Factors. Individuals may be unable to participate in a study because of inadequate financial support (Ballard, Nash, Raiford, and Harrell, 1993; El-Sadr and Capps, 1992). This may result directly from participation, such as when individuals paid on an hourly basis must miss time from work in order to attend appointments at the study and lose the corresponding amount of income, or when they must pay costs such as parking and transportation to get to the study site. The financial cost of participation may also be indirect, as when a prospective participant must allocate a sufficient amount of funds to cover the cost of child care while he or she is at the study site or the costs of meals en route to or back from the study site. These factors may particularly hinder the recruitment of participants with children because of the competing demands for the parent's time and income (Vollmer, Hertert, and Allison, 1992; Smeltzer, 1992).

Legal Difficulties. Some individuals may be reluctant to participate in studies due to the potential legal difficulties that they may face of the information that they divulge in the context of the research becomes known to others. For instance, someone may be concerned that he or she will face disciplinary action as a member of the military should his or her homosexuality or use of recreational drugs become known. Others may be concerned that disclosure of their drug usage or immigration status may result in deportation from the United States.

Methods and Strategies for Recruitment: Voluntariness

Various strategies have been formulated to increase the likelihood of individuals' participation. For instance, the following strategies have been shown to be successful in recruiting African American participants:
1. a commitment to the recruitment of African American participants,
2. efforts to enhance the study's credibility through outreach programs,
3. the involvement of churches and other organizations,
4. publicity campaigns targeting African Americans,
5. attention to patient concerns, such as location and time,
6. the use of incentives,
7. reliance on African American role models,
8. flexibility and willingness to adjust the study design,
9. reliance on lay health workers, and
10. door to door recruitment efforts.

(Shavers-Hornaday, Lynch, Burmeister, and Torner, 1997). Recruitment through individuals' physicians is also common practice.

Although these strategies may work in practice, we must ask whether their usage, and the way they are being used, is appropriate in every instance simply because they may work. Again, we saw that the researchers conducting the Tuskegee experiment understood that recruitment efforts conducted by a trusted individual and the award of a valuable incentive (burial benefits) were likely to succeed both in recruiting individuals initially and in retaining them in the experiment. In formulating strategies to recruit participants, it is helpful to look to some of the international documents for guidance. The Helsinki Declaration provides that:

> When obtaining informed consent for the research project, the physician should be particularly cautious if the subject is in a dependent relationship to him or her or may consent under duress. In that case the informed consent should be obtained by a physician who is not engaged in the investigation and who is completely independent of this official relationship.

With respect to non-therapeutic research involving human subjects (non-clinical biomedical research), the Helsinki Declaration specifically provides:

> The subjects should be volunteers—either healthy persons or patients for whom the experimental design is not related to the patient's illness.

The *International Guidelines for Ethical Review of Epidemiological Studies* (CIOMS, 1991: paragraph 10, at 13) observes that

> [p]rospective subjects may not feel free to refuse requests from those who have power or influence over them. Therefore the identity of the investigator or other person assigned to invite prospective subjects to participate must be made known to them. Investigators are expected to explain to the ethical review committee how they propose to neutralize such apparent influence. It is ethically questionable whether subjects should be recruited from among groups that are unduly influenced by persons in authority over them or by community leaders, if the study can be done with subjects who are not in this category.

Additionally,

> Individuals or communities should not be pressured to participate in a study. However, it can be hard to draw the line between exerting pressure or offering inappropriate inducements and creating legitimate motivation. The benefits of a study, such as increased or new knowledge, are proper inducements. However, when people or communities lack basic health services or money, the prospect of being rewarded by goods, services, or cash payments can induce participation. To determine the ethical propriety of such inducements, they must be assessed in the light of the traditions of the culture. *International Guidelines for Ethical Review of Epidemiological Studies* (CIOMS, 1991: paragraph 11, at 14)

Further,

> Subjects may be paid for inconvenience and time spent, and should be reimbursed for expenses incurred, in connection with their participation in research; they may also receive free medical services. However, the payments should not be so large or the medical services so extensive as to induce prospective subjects to consent to participate in the research against their better judgment ("undue inducement"). All payments, reimbursements, and medical services to be provided to research subjects should be approved by an ethical review committee. (*International Ethical*

Guidelines for Biomedical Research Involving Human Subjects, CIOMS and WHO, 1993: Guideline 4, at 18).

> Risks involved in participation should be acceptable to subjects even in the absence of inducement. It is acceptable to repay incurred expenses, such as for travel. Similarly, promises of compensation and care for damage, injury or loss of income should not be considered inducements. (*International Guidelines for Ethical Review of Epidemiological Studies* CIOMS, 1991: Guideline 12, at 14)

Commentary to Guideline 3 of the *International Guidelines for Biomedical Research Involving Human Subjects* (CIOMS and WHO, 1993: 17) similarly states:

> The investigator should seek to exclude any undue influence on the subject. However, the borderline between justifiable persuasion and undue influence is imprecise....Intimidation in any form invalidates consent.

Accordingly, it appears that mechanisms and strategies for recruitment become problematic where there exists a power differential between the recruiter and the person to be recruited. Differentials in power may be rather easy to detect, as when a community leader is approaching those directly under his authority for their agreement to participate. However, a power differential may also arise due to differences in class and position and may be somewhat more subtle. Consider, again, the Tuskegee study. There, an African-American nurse participated as part of the research team in recruiting individuals to the study. She was a nurse, and therefore held a higher status by virtue of her education, her position, and her income. And, she looked like the participants; perhaps she could be trusted. In fact, it was because she was trusted that the researchers were able to recruit and retain participants into the experiment. Beecher (1970: 289-290) recognized the difficulties inherent in a situation in which a patient's physician is also the researcher attempting to recruit him or her into a study:

> An even greater safeguard for the patient than consent is the presence of an informed, able, conscientious, compassionate, responsible investigator, for it is recognized that patients can, when imperfectly informed, be induced to agree unwisely, to many things....
> A considerable safeguard is to be found in the practice of having at least two physicians involved.... First there is the physician concerned with the care of the patient, his first interest is the patient's welfare; and second, the physician-scientist, whose interest is the sound conduct of the investigation. Perhaps too often a single individual attempts to encompass both roles.

Levine (1992) has suggested that special protections for the prospective participant may be warranted in circumstances where there exists a conflict of interest between a physician-researcher's role as a caregiver of a patient and his or her role as a researcher attempting to recruit an individual into a study.

Even the location at which recruitment is conducted may reflect a power differential or may render the potential participant vulnerable to coercion. For instance, a patient in a mental hospital may not feel free to consent knowing not only that he or she is dependent on the caregivers making the request for participation, but also that he or she may not be entirely free to leave. Whether prisoners are free to consent is always questionable in view of the inherently coercive nature of their environment and the obvious power differential that exists between the prisoners and those in authority. It is for this reason that many countries either prohibit or severely restrict research with prisoners, despite arguments that participation will help relieve prisoners' boredom and provide them with income opportunities (CIOMS and WHO, 1993). United States regulations have, accordingly, restricted research involving prisoners to that which relates to

1. the possible causes, effects, and processes of incarceration, and of criminal behavior...
2. [the] study of prisons as institutional structures or of prisoners as incarcerated persons...
3. conditions particularly affecting prisoners as a class . . .
4. practices...that have the intent and reasonable probability of improving the health or well-being of the [prisoner-participant].

(45 Code of Federal Regulations section 46.304, 1999). For the first two categories of research, the risk can be no more than minimal and no more than an inconvenience to the participants. The latter two categories of research require consultation with experts in appropriate fields. In addition, the IRB must certify (1) that the advantages accruing to the prisoners would not impair their ability to evaluate the risks and benefits of participation, (2) the risks are commensurate with those that would be accepted by nonprisoner participants, (3) the selection procedures are fair, (4) the information provided about the study and the consent forms are clear, (5) participation will not be considered in parole decisions and prisoners are informed about this limitation, and (6) adequate follow-up care is provided (45 Code of Federal Regulations section 46.305, 1999).

As indicated in the international guidelines, recruitment is also problematic where there is an inducement that exceeds what could be considered reasonable under the circumstances. Whether an inducement is reasonable under the circumstances may not be easy to assess. The extent to which a financial payment may influence a decision to participate may be substantial. For instance, a study of medical students' willingness to participate as volunteers in clinical trials found that only 2.9 percent had already volunteered, 39.7 percent said that they would never participate, 32.2 percent said that they would participate for scientific interest and financial reward, and 4.2 percent would participate for the financial reward alone (Bigorra and Baños, 1990). Compared to the medical students, experienced healthy volunteers indicated that the financial reward was the primary reason for participation (90%).

Various factors should be considered in determining the appropriateness of a particular recruitment strategy and of a specific incentive. These are set forth in Table 4 below.

> TABLE 4. Considerations in Formulating Recruitment Strategies
>
> Does the recruitment strategy to be used consider explicit and implicit power differentials that may affect willingness to participate, e.g. a physician-patient relationship, a leader-community resident relationship?
>
> If the recruitment strategy relies on community personnel to assist with recruitment, is the intent behind such reliance to increase participant comfort or to promote one or more forms of deception in research?
>
> Does the incentive to be offered represent a compensation to the participant for costs associated with the participation, such as parking and child care, or does it represent a windfall?
>
> Is the payment to the research participant, whether in money or in kind, so large as to induce him or her to accept undue risks or to volunteer against their better judgment?
>
> What are the individual financial and social circumstances? Does the incentive to be provided represent something that is otherwise unobtainable?
>
> What will be the impact of the individual's receipt of this incentive within his or her family or community? For instance, will it inadvertently create undesirable rivalries or power disparities?
>
> Are there alternative forms of an incentive available that may be less inherently coercive, e.g. coupons for a meal at a local restaurant or for food at a grocery store, rather than a cash payment?

To this point, we have been discussing recruitment in the context of recruiting live participants. However, there are also studies that are conducted in reliance on retrospective record reviews. Consider the following example.

A researcher decides that he or she wishes to investigate potential causes of sudden infant death. He obtains copies of death certificates with this diagnosis of death from the local governmental agency maintaining such records. He then matches these records with the birth records of the children and obtains the names of the physicians listed on the birth and death certificates. He contacts these local physicians to ask for copies of the children's medical records, if any, and the mother's records during her pregnancy. The mothers are not contacted for their consent to view their records because the researcher is concerned that, if they are contacted, they will become traumatized by the recollection of the event and will require referrals for counseling or supportive services. What are the ethical implications of such "recruitment"?

The *International Guidelines for the Ethical Review of Epidemiological Studies* (CIOMS, 1991: 12) provides in the third Guideline the following counsel for such situations:

An ethical issue may arise when occupational records, medical records, tissue samples, etc. are used for a purpose for which consent was not given, although the study threatens no harm. Individuals or their public representatives should normally be told that their data might be used in epidemiological studies, and what means of protecting confidentiality are provided....

This situation can be analogized to that of deception, discussed in chapter 3. However, instead of being given misinformation, the participants are given no information. Some researchers might argue that no harm is being done. After all, if you don't know something, how can it hurt you? However, several issues remain unanswered with this response. First, a study such as this one does not recognize the ability of individuals to control the distribution of information about themselves. Second, the conclusion that no harm has been done views the research solely from the perspective of physical harm and from the researcher's vantage point. Perhaps the women believed that their medical records could not be accessed without their consent (see Roach, Jr. and Aspen Health Law Center, 1994). An intrusion into their records then represents a violation of their privacy. One might argue, again, that if they don't know of the intrusion, it can't be a violation. However, if someone were to enter a home without the owner's consent or knowledge, the entrance is not any less of an intrusion because the owner has not yet been made aware of its occurrence. Finally, such practices have the potential to diminish respect for science and to fuel conspiracy theories.

Methods and Strategies for Recruitment: Justice

Recruitment also, however, raises issues of distributive justice: whether there is to be an equal sharing of the burdens and the benefits of research. Although the research design may not specifically exclude persons of specified groups, the manner in which recruitment is conducted may, in fact, lessen the likelihood that members of various groups will participate. For instance, assume that a behavioral intervention trial to reduce risk behavior for the transmission of an infectious disease is to be conducted. Interviews will take place only between the hours of 9 and 5 and only at the investigator's office, where there are no provisions for child care. Individuals will be recruited through health care providers. This recruitment scheme is likely to result in the relative exclusion of individuals who are working on an hourly basis, because they will lose money while away from work; of individuals who reside or work a long distance from the office, due to the length of time and possible logistical difficulties associated with travel; and women with small children and restricted incomes, who have no alternatives for child care for their children. Individuals without regular health care providers may also be excluded; many of these individuals may be of lower economic status and/or may lack health insurance coverage. Depending upon the demographic features of a community, these restrictions may, in turn, disparately impact various ethnic or racial communities. Conversely, we have seen in chapter 1 how members of some groups were recruited into research specifically because of their characteristics: they were easily available, relatively compliant, and their recruitment was relatively inexpensive. One writer concluded after his review of the literature that

> Clinical studies in the past have been conducted more frequently among impoverished minorities than among the privileged American classes. The poor were enrolled as subjects in medical investigations because it was convenient (researchers were located in the teaching hospitals used by very poor patients), and because of gross insensitivity to the unfairness of the custom. (Silverman, 1989: 9).

Various international and national documents address this issue. The *International Ethical Guidelines for Biomedical Research Involving Human Subjects* (CIOMS and WHO, 1993: 29) speaks to the equitable distribution of burdens and benefits and the recruitment process:

> Individuals or communities to be invited to be subjects of research should be selected in such a way that the burdens and benefits of the research will be equitably distributed. Special justification is required for inviting vulnerable individuals and, if they are selected, the means of protecting their rights and welfare must be particularly strictly applied.

In its explanation of ethical principles to be applied in epidemiological studies, the *International Guidelines for Ethical Review of Epidemiological Studies* (CIOMS, 1991: 11) explains that

> *Justice* requires that cases considered to be alike be treated alike and the cases considered to be different be treated in ways that acknowledge the difference. When the principle of justice is applied to dependent or vulnerable subjects, its main concern is with rules of *distributive justice*. Studies should be designed to obtain knowledge that benefits the class of persons of which the subjects are representative: the class of persons bearing the burden should receive an appropriate benefit, and the class primarily intended to benefit should bear a fair proportion of the risks and burdens of the study.
>
> The rules of distributive justice are applicable within and among communities. Weaker members of communities should not bear disproportionate burdens of studies from which all members of the community are intended to benefit, and more dependent communities and countries should not bear disproportionate burdens of studies from which all communities or countries are intended to benefit. (Italics in original.)

Guideline 42 of the same document speaks to equity in the selection of research participants:

> Epidemiological studies are intended to benefit populations, but individual subjects are expected to accept any risks associated with studies. When research is intended to benefit mostly the better off or healthier members of a population, it is particularly important in selecting subjects to avoid inequity on the basis of age, socioeconomic status, disability, or other variables. Potential benefits and harm should be distributed equitably within and among communities that differ on grounds of age, gender, race, or culture, or otherwise. (CIOMS, 1991: 21).

The National Commission for the Protection of Human Subjects of Biomedical and Behavioral Research (1978: 9-10) advised that:

> the selection of research subjects needs to be scrutinized in order to determine whether some classes (e.g., welfare patients, particular racial and ethnic minorities, or persons confined to institutions) are being systematically selected simply because of their easy availability, their compromised position, or their manipulability, rather than for reasons directly related to the problem being studies. Finally, whenever research supported by public funds leads to the development of therapeutic devices and procedures, justice demands both that these not provide advantages only to those who can afford them and that such research should not unduly involve persons from groups unlikely to be among the beneficiaries of subsequent applications of the research.

The *Institutional Review Board Guidebook* (Penslar, no date) says of the principle of justice and recruitment:

> [S]ubjects should not be selected either because they are favored by the researcher or because they are held in disdain (*e.g.*, involving "undesirable persons in risky research). Further, "social justice" indicates an "order of preference in the selection of classes of subjects (*e.g.*, adults before children) and that some classes of subjects, (e.g., the institutionalized mentally inform or prisoners) may be involved as research subjects, if at all, only on certain conditions."

Explaining the Study

Information

The international documents to which we have repeatedly referred provide guidance as to the extent of the information that is required to be provided to a prospective participant in connection with a solicitation of his or her participation in research. The Helsinki Declaration provides:

> In any research on human beings, each potential subject must be adequately informed of the aims, methods, anticipated benefits and potential hazards of the study and the discomfort it may entail. He or she should be informed that he or she is at liberty to abstain from participation in the study and he or she is free to withdraw his or her consent to participation at any time. The physician should then obtain the subject's freely-given informed consent, preferably in writing.

Although the Helsinki Declaration states a preference for a written statement of informed consent, it is critical to remember that the written document serves as evidence of the informed consent process; it is not the consent itself. Further, documentation may be "an alien concept in many cultures" (Gostin, 1991: 193) or it may be regarded with fear and suspicion due to the historical and social contexts in which the documentation was utilized in the past (Loue, Okello, and Kawuma, 1996).

Guideline 2 of the *International Ethical Guidelines for Biomedical Research Involving Human Subjects* (CIOMS, 1993) enumerates the elements of the information that are to be provided to each prospective participant as part of the informed consent process: that it is research, the aims and methods of the research, the expected duration of the individual's participation, the potential benefits to the individual or to others, any foreseeable risks or discomforts, any alternative procedures or treatments that could be as advantageous to the individual as that under examination, the extent to which confidentiality will be maintained, the extent of the investigator's responsibility to provide medical services, that therapy will be provided at no cost for specified research-related injuries, whether there is a provision for the compensation of the participant and/or his or her family in the event of research-related disability or death, and that the individual may withdraw at any time without the loss of any benefits to which he or she would otherwise be entitled. Guideline 3 imposes on the investigator an affirmative obligation to communicate all of the information necessary for informed consent, to give the prospective participant an opportunity to as questions and receive answers, to exclude the possibility of unjustified deception or undue influence or intimidation, to request consent only after the individual has received adequate information and has had sufficient time to consider participation, to obtain a signed form evidencing that consent, where possible, and to renew the informed consent if there are material changes in conditions or procedures. This last requirement of consent renewal is discussed below in the context of continuing consent.

The *International Guidelines for Ethical Review of Epidemiological Studies* (1991: Guideline 9, 13) recognizes both the need to provide the prospective participant with information and the possibility of deception or the selective disclosure of information. The Guideline cautions:

> In epidemiology, an acceptable study technique involves selective disclosure of information, which seems to conflict with the principle of informed consent. For certain epidemiological studies, non-disclosure is permissible, even essential, so as not to influence the spontaneous conduct under investigation, and to avoid

obtaining responses that the respondent might give in order to please the questioner. Selective disclosure may be benign and ethically permissible, provided that it does not induce subjects to do what they would not otherwise consent to do. An ethical review committee may permit disclosure of only selected information when this course is justified.

Although the *International Guidelines for the Ethical Review of Biomedical Research Involving Human Subjects* (CIOMS, 1991: 17) also acknowledges that deception may be employed as a research strategy, it suggests restrictions on its use:

Deception of the subject is not permissible in research projects that carry more than a minimal risk of harm to the subject. When deception is indispensable to the methods of an experiment, the investigator must demonstrate to the ethical review committee that no other research method would suffice; that significant advances could result from the research; and that nothing has been withheld that, if divulged, would cause a reasonable person to refuse to participate. The ethical review committee with the investigator should determine whether and how deceived subjects should be informed of the deception upon completion of the research. Such informing, commonly called "debriefing," ordinarily entails explaining the reasons for the deception. A subject who disapproves of having been deceived is ordinarily offered an opportunity to refuse to allow the investigator to use information obtained from studying the subject.

Accordingly, various regulations set forth the nature of the information that must be provided to a research participant. The Food and Drug Administration, for instance, requires that the following elements be included:
1. a statement that the study involves research
2. an explanation of the purposes of the research
3. the expected duration of the participant's involvement
4. a description of the procedures to be followed
5. the identification of any procedures that are experimental
6. a description of any reasonably foreseeable risks or discomforts
7. a description of any benefits to the subject or to others that are reasonably expected,
8. a disclosure of appropriate alternative procedures or treatments
9. a statement describing the extent to which confidentiality will be maintained and an indication that the FDA may inspect the records
10. where the research involves more than minimal risk, an explanation as to whether any compensation or medical treatment is available of an injury should occur and where further information may be obtained,
11. an explanation of who to contact for further information and in case of a research-related injury
12. a statement that participation is voluntary, that refusal to participate will not involve the loss of any benefit to which the individual is

otherwise entitled, and that the individual may discontinue his or her participation at any time without penalty or loss of benefits to which he or she is otherwise entitled. (21 Code of Federal Regulations section 50.20, 1999).

Additional elements may be included, such as the possibility of any costs to the participant as a result of his or her participation, the potential for risks to the fetus, the circumstances under which the investigator may terminate the participant's involvement without his or her consent, the consequences of the participant's decision to withdraw from the study and procedures for such termination, a statement that the investigator will provide to the participants information that develops during the course of the study that may relate to the individual's willingness to continue his or her participation, and the approximate number of individuals involved in the study (21 Code of Federal Regulations, section 50.20, 1999).

Despite the guidance offered by the international documents and, in the United States, the mandate imposed by the federal regulations, questions frequently arise about the extent to which details of the study must be provided to the participants (Dal-Re, 1992; Tobias, 1988). Researchers may be concerned, for instance, that full disclosure will frighten potential participants and discourage their enrollment, resulting in the prolongation of the planned recruitment period or the abandonment of the research (Lara and de la Fuente, 1990; Thong and Harth, 1991). It is important, however, that the individual be provided with sufficient information so that he or she can evaluate how participation will impact on his or her life. The failure to provide sufficient detail about the research at the commencement of an individual's participation may facilitate his or her enrollment, but may result in the individual's later withdrawal from the study or refusal to adhere to study protocol.

Understanding and Capacity

In addition to understanding, informed consent requires that the prospective participant both understand the information that he or she is given and that he or she have capacity to consent. In some cases, the ability to understand may be directly related to capacity. For instance, a 17-year old adolescent may understand fully the information presented to him or to her, but lacks the legal capacity to consent by virtue of his or her age. In contrast, a two-year old child has neither the ability to understand nor the legal capacity to give consent, while a schizophrenic man may be legally competent and yet lack both capacity and understanding. These elements are reflected in both the *International Ethical Guidelines for Biomedical Research Involving Human Subjects* and the Helsinki Declaration.

> Informing the subject must not be simply a ritual recitation of the contents of a form. Rather, the investigator must convey the information in words that suit the individual's level of understanding. The investigator must bear in mind that ability to understand the information necessary to give informed consent depends on the individual's maturity, intelligence, education and rationality....

The investigator must then ensure that the prospective subject has adequately understood the information. This obligation is the more serious as risk to the subject increases. In some instances the investigator might administer an oral or written test to check whether the information has been adequately understood. (CIOMS and WHO, 1993: Commentary to Guideline 2, 14-15)

In case of legal incompetence, informed consent should be obtained from the legal guardian in accordance with national legislation. Where physical or mental incapacity makes it impossible to obtain informed consent, or when the subject is a minor, permission from the responsible relative replaces that of the subject in accordance with national legislation....
Whenever the minor child is in fact able to give a consent, the minor's consent must be obtained in addition to the consent of the minor's legal guardian. (Helsinki Declaration)

Readability. Numerous rules have been formulated and strategies devised to increase the likelihood that participants will understand the information that is presented. For instance, the Office for the Protection from Research Risks (OPRR) has strongly suggested that participants who do not speak English be provided with a consent document that is written in the language that is best understandable to them (Director, Division of Human Subject Protections, 1995). (See chapter 5 for a discussion of OPRR and its functions.) Comprehension of the written form, which again represents but one component of the informed consent process, will be facilitated if it is written at an appropriate readability level. It has been estimated that one out of five Americans is functionally illiterate and lacks the reading and writing skills needed for daily activities; their reading ability is at or below the fifth grade reading level (Doak and Doak, 1987). Although the mean educational level of Americans is approximately 12.6 years of school (United States Department of Education, 1986), their reading level is often three to four grade levels lower than the stated years of schooling (Boyd and Feldman, 1984; Doak and Doak, 1980). Readability level may be measured by using instruments such as the Fry Readability Scale or the Flesch Readability Formula (Silva and Sorrell, 1988), both of which determine the readability level by relying on computations involving the number of sentences and syllables per designated selection. Many researchers have recommended that the reading level of an informed consent form be no higher that the eighth grade. The lower the readability level is, the more likely that the prospective participant will be able to comprehend the information presented (LoVerde, Prochazka, and Byyny, 1989; Young, Hooker, and Freeberg, 1990). Understanding will also be increased through the use of simple sentences, the repetition of nouns rather than pronouns, the avoidance of metaphors, the avoidance of the passive voice, and the avoidance of the subjective mood.

Readability, however, is only one of several factors that may affect an individual's comprehension (Silva and Sorrell, 1988). Confinement to bed has been found to affect comprehension negatively (Cassileth, Zupkis, Sutton-Smith, and March, 1980). Reliance on nonmedical personnel or a third party to present and review information for informed consent has been determined to increase comprehension (Benson, Roth, Appelbaum, Lidz, and Winslade, 1988; Muss, White, Michielutte et al., 1979). Understanding may be increased if the prospective participant is given sufficient time to understand the information prior to signing the consent document (Lavelle-Jones, Byrne, Rice, and Cuschieri, 1993; Moorow, Gootnick, and Schmale, 1978; Tankanow, Sweet, and Weiskopf, 1992). A simple, clear, and concise manner of presentation on the written consent form has also been found to enhance comprehension (Epstein and Lasgana, 1969; Simel and Feussner, 1992).

Additional modifications of presentation may be necessary to facilitate comprehension, particularly where the individual has experienced loss in the ability to organize and integrate information (Peterson, Clancy, Champion, and McLarty, 192). Organizational modifications include varying the size of type or the spacing of information or using advance organizers (Taub, 1986), using a multicomponent program that includes both written materials and other audiovisual aids (DCCT Research Group, 1989), or the use of graphics and summary declarative statements (Peterson, Clancy, Champion, and McLarty, 1992). The use of a video as a way of explaining the research has been found to be particularly helpful with prospective participants in psychiatric research (Benson et al., 1988).

<u>Decision-Making Capacity and the Cognitively Impaired</u>. Limited decision making capacity refers to a broad spectrum of individuals, encompassing those who are severely retarded and do not and will not have capacity to make decisions, those who may be temporarily impaired due to shock, and those whose decision making capacity may fluctuate, such as individuals with schizophrenia (Sunderland and Dukoff, 1997). The involvement of certain groups of persons in research is likely to raise concerns about understanding and capacity due to various actual and/or attributed characteristics of those groups' members. For instance, we saw in chapter 3 that individuals who are mentally ill are not *per se* incapable of decision making as the result of their illness. However, as a class, they are often treated, both legally and socially, as if they are incapable. This obviates the need for a court to require an assessment of each individual's capabilities.

Conversely, individuals who lack capacity to provide informed consent may be treated as if they have such capacity and are, consequently, vulnerable to abuse. As an example, recent research involving medication withdrawal from persons with schizophrenia of recent onset at the University of California at Los Angeles was cited by the Office for Protection from Research Risks (1994) of the United States Department of Health and Human Services for deficiencies in obtaining informed consent (see Appelbaum, 1996; Katz, 1993). Such studies, conducted to determine which patients will do well without medication and how to minimize dosages, involve the random withdrawal of patients

> from their ongoing medication to either a standard drug or to a placebo so that the treatment was double blind. Whether or not the patient relapsed and how long it took until relapse, were the reported outcomes….Typically the withdrawal studies were of longer duration, three, four, or even nine months…. (Schooler and Levine, 1983, cited in Shamoo and Keay, 1996: 374).

A recent study of 41 relapse studies conducted in the United States between 1966 and 1993 found that authors in 15 of the studies did not report getting informed consent, in 23 the participants signed informed consent forms, and in 3, next of kin were authorized to give consent. However, in 39 of the studies there was no mention of whether the participants had capacity to give consent or whether they had been assessed for capacity to make decisions (Shamoo and Keay, 1996). These studies were conducted despite the fact that there was evidence as early as 1975 indicating that appropriate drug treatment reduces relapse occurrence (Davis, 1975, 1985; Schooler and Levine, 1983). Various other studies induced relapse of schizophrenia through a drug "challenge" such as amphetamines or L-dopa (Angrist, Peselow, Rubenstein, Wolkin, and Rotrosen, 1985; Davidson, Keefe, Mohs, Siever, Losonczy, Horvath, and Davis, 1987; Van Kammen, Docherty, and Bunney, 1982). In 1996, a New York appellate court halted various studies being conducted in public psychiatric hospitals in which surrogates had provided consent on behalf of incompetent patients to participate in research involving greater than minimal risk with no direct therapeutic benefit to the participants (*T.D. v. New York Office of Mental Health*, 1996).

Various contextual factors may heighten the potential vulnerability of an individual with decision making impairment. Such conditions include institutionalization, such as in a mental hospital or nursing home, due to the presumed *per se* restrictive and/or coercive nature of such a setting (American College of Physicians, 1989; Melnick, Dubler, Weisbard, and Butler, 1985; Sachs, Rhymes, and Cassel, 1993; see Klerman, 1977) and conflict or stress within the individual's family, so that participation in research represents a period of respite for the caregiver(s), who is/are too willing to consent to that participation (Keyserlingk, Glass, Kogan, and Gauthier, 1995).

The Nuremberg Code provides that "[p]roper preparations should be made and adequate facilities provided to protect the experimental subject even against remote possibilities of injury, disability, or death." The *International Ethical Guidelines for Biomedical Research Involving Human Subjects* (CIOMS and WHO, 1993) recognize that some individuals with mental or behavioral disorders may not be capable of giving informed consent. The Guidelines specifically provide that

> The willing cooperation of subjects should be sought to the extent that their mental state permits, and any objection on their part to taking part in any non-clinical research should always be respected. When an investigational intervention is intended to be of therapeutic benefit to a subject, the subject's objection should always be respected unless there is a reasonable medical alternative and local law permits overriding the objection. (Commentary to Guideline 6, at 22-23)

As such, the Guideline recognizes the autonomy of the mentally ill person to a limited degree. With respect to experimental therapeutic interventions, however, the Guideline presumes that a mentally ill individual lacks the capacity or understanding to choose between no intervention and the consequences of such versus the risks and benefits of the proposed intervention. It is critical to recognize here that we are not referring to standard, approved medical treatment, but instead to an experiment in which the individual refuses to participate. Pellegrino (1992: 368) has cautioned:

> The safe rule in [clinical research] is to favor beneficence over scientific rigor when the two seem to be in conflict or in doubt. The possible loss of knowledge cannot outweigh the possibility of harm to the subject even if the utilization calculus indicates great benefit to many and harm to only a few.

Whether the individual lacks decision making capacity requires individualized assessment. Several authors have suggested that, in the context of research with individuals with mental impairment, such as those with Alzheimer's disease, an assessment of understanding requires a determination of whether the individual understands the nature of the research, the nature of his or her participation in the research, the consequences in his or her life of (non)participation, the fact that the proposed activity or intervention is research, the intended therapeutic benefit or lack thereof to the individual of the research, the right to refuse to participate or to withdraw at any time, the likelihood that the participant will become incompetent during the course of the research, as in the case of advancing Alzheimer's disease, and the fact that a decision to withdraw will not adversely affect the care being received (Keyserlingk, Glass, Kogan, and Gauthier, 1995). Shamoo and Irving (1997:38) have noted, though, the difficulty inherent in any assessment of capacity because "it is not unusual for an incompetent [mentally ill] person to appear to be competent when they are not; many learn how to play that game."

The legal response to the inclusion of individuals with impaired decision making capacity in research has varied across countries. Sweden prohibits clinical trials involving drugs with individuals suffering from psychiatric illness unless the trial relates to the treatment of the mental illness. France appears to permit nontherapeutic experimentation with the mentally impaired if there is no serious health risk to the participant, the research relates directly to mental illness or the handicap, and the experiment cannot be conducted with other participants (Baudouin, 1990).

Current United States regulations do not specifically address research involving persons who are cognitively impaired. The National Institutes of Health (no date) has suggested the following points to consider in conducting such research:

1. Individuals with cognitive impairments may find it difficult to understand the researcher-physician's multiple roles. The consent process must clearly explain the difference between treatment and research, as well as researcher and clinician.
2. The IRB should include one voting member independent of the research with experience in working with individuals of questionable capacity and/or additional members from the community.
3. The individual's capacity must be adequately assessed.
4. The safeguards for the participants should increase as the impairment increases in its severity.

5. Ongoing educational efforts should be conducted to increase participants' understanding.
6. Additional safeguards may include reliance on an independent monitor when greater than minimal risk is involved, reliance on a surrogate using substituted judgment (deciding in a manner that the individual would have decided him- or herself if he or she were able to decide), reliance on assent of the prospective participant, utilization of advance research directives, use of educational aids and strategies, and incorporation of waiting periods into the informed consent process for the provision of information about the study incrementally and to permit sufficient time to process the information (see Expert Panel Report to the National Institutes of Health, 1998).

Various researchers have argued strongly that our approach to the participation of mentally ill persons in research is in desperate need of reform (Capron, 1999; Moreno, Caplan, Wolpe, and Members of the Project on Informed Consent, 1998). First, by not setting higher requirements for informed consent, patients recruited for research may become victims to the "therapeutic misconception," whereby the intervention is construed as advantageous because it is an intervention, even where there is little probability of benefit (Capron, 1999). Second, researchers may forget that reliable, generalizable knowledge must be derived from a scientific approach, and cannot result from the accumulated observations of interventions in individual cases. Accordingly, Capron (1999:1432) has advocated reliance on "an independent, qualified professional [to] assess the potential [participant's] capacity to consent to participate in any protocol that involves greater than minimal risk." Although the National Bioethics Advisory Commission has not adopted Capron's recommendation, it has concluded that one of three conditions must be met as a prerequisite to the enrollment of a participant: (1) informed consent given while the participant had decision making capacity, (2) prospective consent for specified types of research given while the individual had decision making capacity, or (3) permission from a legally authorized representative chosen by the individual or from a concerned relative or friend who is both available to monitor the individual's involvement and who will premise decisions on what the individual would have chosen if he or she were able to make a decision (National Bioethics Advisory Commission, 1998). At least one other research group has advocated that federal regulations be amended to require that investigators working with cognitively impaired research participants include in their protocols "[a] written description concerning the expected degree of impairment of the subjects, as well as the means by which it will be determined that they have lost the ability to make decisions regarding research participation . . . . (Moreno, Caplan, Wolpe, and Members of the Project on Informed Consent, 1998: 1953).

<u>Research with Children</u>. Children are, in general, unable to give informed consent to participate in research for a number of reasons. First, they may lack the maturity to understand what research means and the nature of their participation. Second, although the specific age of majority differs across societies, children are generally deemed to lack legal capacity to consent. This lack of understanding and the lack of capacity, which flows from a recognition of children's inability to understand, combine to make children especially vulnerable as participants in research.

We saw in chapter 1 the unfortunate consequences that can attend that vulnerability absent adequate controls on experimentation: the injection of sterilized gelatin into several children, resulting in their collapse (Abt, 1903); the injection of tuberculin solution into young, orphaned children, causing eye lesions and inflammation (Belais, 1910; Hammill, Carpenter, and Cope, 1908); the administration of oatmeal laced with radioactive calcium or iron to the boys of the Fernald School (Welsome, 1999); and the deliberate infection of mentally retarded institutionalized children at Willowbrook with hepatitis (Beecher, 1970). Periodic protests apparently did little to impede such efforts. Bercovici, a social worker and journalist, castigated in 1921 a researcher for his deliberate inducement of scurvy in institutionalized infants:

> No devotion to science, no thought of the greater good to the greater number, can for an instant justify the experimenting on helpless infants, children particularly abandoned by fate and entrusted to the community for their safeguarding. Voluntary consent by adults should, of course, be the sine qua non of scientific experimentation (Bercovici, 1921: 913

Yet, we see that essentially uncontrolled experimentation with children continued well beyond that date.

Notwithstanding this misuse of children, there are clearly valid reasons to involve them as participants in research. Children differ from adults, both anatomically and physiologically. As discussed in chapter 3, certain diseases may be unique to children, such as newborn respiratory distress syndrome and neuroblastoma. Other diseases, such as mumps and measles, are more likely to occur in during childhood than during the adult years. Absent research with children, it may not be possible to discover new treatments and procedures that could prevent or ameliorate the effects of such diseases. However, reliance on children in research raises numerous issues that must be addressed. To what extent can/should the child be involved in the decision to participate? What information should be disclosed to the child and in what manner? To what extent is the child entitled to privacy and confidentiality, e.g., from disclosure of information to his or her parents? What is the level of risk to the child? What is the resulting benefit—medical or otherwise—to the child? To others?

Accordingly, the ethical basis for the participation of children in research remains controversial. Ramsey (1989) has maintained that experimentation involving human beings which is not for their benefit cannot be performed without their informed consent. Since children are incapable of giving informed consent, he argues that experimentation on young children is justified only if it is the best means to effectuate the child's recovery from a disease or condition or if it is intended to protect the child from some greater risk. Redmon (1986: 81), however, has argued that children can participate in research which is not intended to be beneficial to them

> if we can reasonably expect this child to 'identify'...with the goals of the research when she is an adult, and that the identification will be strong enough to outweigh the harm of the knowledge of being

used by her parents, and if the child (if old enough) assents, and if the possibility of harm is slight ('minimal risk')....

Guideline 5 of the *International Ethical Guidelines for Biomedical Research Involving Human Subjects* (CIOMS and WHO, 1993: 20) provides some guidance:

> Before undertaking research involving children, the investigator must ensure that:
> --children will not be involved in research that might equally well be carried out with adults;
> --the purpose of the research is to obtain knowledge relevant to the health needs of children;
> --a parent or legal guardian of each child has given proxy consent;
> --the consent of each child has been obtained to the extent of the child's capabilities;
> --the child's refusal to participate in the research must always be respected unless according to the research protocol the child would receive therapy for which there is no medically acceptable alternative;
> --the risk resented by interventions not intended to benefit the individuals child-subject is low and commensurate with the importance of the knowledge to be gained; and
> --interventions that are intended to provide therapeutic benefit are likely to be at least as advantageous to the individual child-subject as any available alternative.

Current federal regulations also incorporate special protections for children participating in research. The requirements relating to parental permission, the child's assent, and ultimate approval of the proposed research are directly linked to the level of risk and the anticipated benefits. Research is classified into four categories.

Category 1: Research involving not more than minimal risk may be approved if adequate provisions are made for obtaining the child's assent and the parent's or guardian's permission. (45 Code of Federal Regulations section 46.404, 1999). "Minimal risk" means that the risks of harm anticipated in the proposed research are not greater, considering probability and magnitude, than those ordinarily encountered in daily life or during the performance of routine physical or psychological examinations or tests. (45 Code of Federal Regulations section 46.102, 1999).

Category 2: Research involving greater than minimal risk, but with the possibility of yielding direct benefit to the child. Approval requires that adequate provisions have been made for obtaining the child's assent and the parent's or guardian's permission and that the IRB find that (a) the risk is justified by the anticipated benefit to the child and (b) the risk/benefit ratio is at least as favorable as other alternative approaches available to the child. (45 Code of Federal Regulations section 46.405, 1999).

Category 3: Research which involves greater than minimal risk and will not yield direct benefit to the child, but will most likely produce generalizable knowledge about the child's disease or condition. Approval requires that the IRB find that the risk represents a minor increase over minimal risk, that the procedures are "reasonably commensurate" with those inherent in the child's situation, and that the research is likely to yield generalizable knowledge of "vital importance" to the understanding and amelioration of the disease or condition under investigation. Unlike research in categories 1 and 2, research in this category requires the child's assent and the permission of both parents. (45 Code of Federal Regulations section 46.406, 1999).

Category 4: Research not encompassed by categories 1 through 3. Approval requires a finding by the Secretary of HHS, after consultation with an expert panel, that the research provides a "reasonable opportunity" to understand, prevent, or ameliorate a "serious problem" that affects the health and welfare of children and the research will be conducted in accordance with "sound ethical principles." (45 Code of Federal regulations section 46.407, 1999).

The concept of assent is critical in affording children a voice in what is to happen to them. It serves to operationalize, to the extent possible with children, the principle of respect for persons and the recognition of an individual as an autonomous agent. In addition, the process serves to provide the child with information (Leikin, 1993), and may serve as a mechanism to provide moral training (Bartholme, 1976). Gaylin (1982: 49) provides as an example the statement of a father who explained, in ordering his young son to give a small sample of his blood for research over the child's objection that it would hurt, that it was

> His moral obligation to teach his child that there are certain things that one does, even if it causes a small amount of pain, to the service or benefit of others. This is my child. I am less concerned with the research involved than with the kind of boy that I was raising. I'll be damned if I was going to allow my child, because of some idiotic concept of children's rights, to assume that he was entitled to be a selfish, narcissistic little bastard."

What assent means, and when a child is capable of assent, is unclear. The regulations define assent as "a child's affirmative agreement to participate in research. Mere failure to object, should not, absent affirmative agreement, be construed as assent." (45 Code of Federal Regulations section 46.402, 1999). The IRB in reviewing the protocol is charged with the responsibility of evaluating the children's ability to assent, based on their age, level of maturity, and psychological state (45 Code of Federal Regulations section 46.408, 1999). Assent is not, however, required if the IRB finds that the children's ability to assent is so limited that they cannot be consulted or the procedure or the treatment contemplated is likely to be of direct benefit to the child participants and is available only in the context of the research (45 Code of Federal Regulations section 46.408, 1999). This seems to suggest that, at least in the context of research that may yield some direct benefit, it is permissible to seek assent or to override the child's veto (Leikin, 1993). The regulations do not, however, address whether the child's veto can be overridden in research that is nontherapeutic.

The Committee on Bioethics of the Academy of Pediatrics has attempted to delineate more specifically the meaning of assent in both the clinical care and research contexts. Accordingly, assent must include, at a minimum,

1. helping the patient achieve a developmentally appropriate awareness of the nature of his or her condition,
2. telling the patient what he or she can expect with tests and treatment(s),
3. making a clinical assessment of the patient's understanding of the situation and the factors influencing how he or she is responding (including whether there is inappropriate pressure to accept testing or therapy), and
4. soliciting an expression of the patient's willingness to accept the proposed care. (Committee on Bioethics, 1998: 59).

It is unclear, however, to what extent children can either understand or reason about research. One research group found that children of all ages who were on therapeutic research protocols were able to understand various concrete elements, including the duration of the treatment, what was required of them, the potential benefits resulting from their participation, and the facts that their participation was voluntary and that they could ask questions (Sussman, Dorn, and Fletcher, 1992). However, other studies indicate that children may not understand either that they are participating in research (Schwartz, 1972; Sussman, Dorn, and Fletcher, 1992) or the scientific purposes underlying the research (Ondrusek, Abramovitch, and Koren, 1992; Sussman, Dorn, and Fletcher, 1992). Additionally, they may be unable to recall the benefits to others that may derive from the research, the alternative treatments available, the risks attending participation (Schwartz, 1972), or that they may withdraw from participation (Ondrusek, Abramovitz, and Koren, 1992; Sussman, Dorn, and Fletcher, 1992). Various other studies, though, have suggested that children are able to identify the risks and benefits and incoporate these elements into their decisionmaking processes (Kayser-Boyd, Adelman, Taylor, and Nelson, 1986; Keith-Spiegel and Maas, 1981; Lewis, Lewis, and Ifekwunigue, 1978; Weithorn and Campbell, 1982). Children's understanding of their rights, such as the right to withdraw, appears to relate to their level of moral judgment. Melton (1980) has identified three developmental stages: level 1, at which young children believe that they have rights only if adults so permit; level 2, at which children believe that rights based on fairness are linked to being good or being nice; and level 3, where children are able to conceptualize of rights based on abstract universal principles, such as the right to privacy.

It is not surprising, in view of the complexities of conducting research with children, that only a small proportion of drugs and biological products that are marketed in the United States have been investigated in clinical trials with children. Most marketed drugs are not labeled for use in children. Because of these deficiencies, the Food and Drug Administration proposed new regulations that would ensure that manufacturers of prescription drugs test their drugs' effects on children if the medications will have a clinically significant use in children (National Institutes of Health, 1998). NIH similarly adopted a policy, effective October 1, 1998, which provides that children (under the age of 21) must be included in research protocols unless there are scientific or ethical reasons to exclude them. This policy reflects a choice made between two undesirable outcomes:

> Society may choose to forbid drug evaluation in pregnant women and children. This choice would certainly reduce the risk of damaging individuals through research. However, this would maximize the possibility of random disaster resulting from the use of inadequately investigated drugs. In the final analysis it seems safe to predict that more individuals would be damaged; however, the damage would be distributed randomly rather than imposed upon preselected individuals. (Mirkin, 1975: 110)

Various arguments have been made to support parents' interest in consenting to or withholding consent to their children's participation in research. First, children lack the legal capacity, in most cases, to decide themselves and their parents or guardians are presumed to understand their children's needs and have their children's best interests in mind. Second, parents or guardians should have some control over the decision to participate because they will bear the consequences of that decision. Third, parents should have discretion in deciding what values to impart to their children. Fourth, the family as an institution is entitled to some degree of privacy. Finally, children would most likely want their parents to make such decisions for them (Brock, 1994; Committee on Bioethics, 1998; Melton, 1989).

Notwithstanding this utopian image of the parent-child relationship, it is clear that some parents breach their obligations towards their children, often through abuse and/or neglect (Committee on Bioethics, 1998), resulting in the placement of the children with foster parents, who may lack the legal right to consent to the children's participation in research. As of 1990, seven states had formal policies relating to the participation of foster children in clinical trials and five states reported having a mechanism by which foster children could be enrolled in clinical trials. Of those twelve states, four required the consent of the child's biological parents as a prerequisite to participation (Martin and Sacks, 1990). Other methods of obtaining consent may be possible where the biological parent is unavailable or unable to give consent. These include a case-by-case determination; an award of medical guardianship to the foster parents; and the submission of protocols to a central review board, which will also review the enrollment of children on a case-by-case basis (Levine, 1991).

The assessment of what, in practice constitutes minimal risk, and the evaluation of the risk-benefit ratio are not unproblematic (see Freedman, Fuks, and Weijer, 1993). For instance, one research group surveyed chairpersons of pediatric departments and directors of pediatric clinical research units in the United States to ascertain their appraisal of the risk level associated with tympanocentesis (puncturing of the eardrum). Fourteen percent believed that it represented minimal risk, 46 percent classified it as representing a minor increment over minimal risk, and 40 percent believed that it reflected more than a minor increase over minimal risk (Janosky and Starfield, 1981).

Too, the evaluation of the risk of harmful outcome in research procedures may depend to some extent on the child's stage of development. For instance, younger children are more likely to be adversely affected by a procedure that removes them from a familiar caregiver, while older children are more likely to be disturbed by procedures that affect function and distort body image (Wender, 1994).

The assessment of the risk-benefit ratio may be even more difficult when the children to be enrolled in a study are quite ill. Ackerman (1980:2) has pointed out, for instance, that attempts to expedite pediatric research to enhance the welfare of HIV-infected children in general may ultimately compromise the welfare of the individual children participating in such drug trials:

> First, children recruited for phase I trials will be in the later stages of the illness, since they are not usually eligible until phase II or phase III drugs have proven ineffective or unacceptably toxic in their treatment. As a result, they will have incurred a substantial burden of prior suffering and may be significantly debilitated. Second, many subjects in phase I trials do not receive potentially therapeutic doses of the drugs being studied, because the increments in dosing for consecutive groups of subjects usually begin with a very low conservative dose. Third, exposure to unexpected toxicities and additional monitoring procedures may compound the suffering....Finally, the harm/benefit ratio of participating in a phase I trial must be compared to the alternative management strategy of using only measures that will enhance the comfort of the patient.

Exercise

You are a health researcher from the United States. The country of Technovy has experienced a decline in its birth rate, so that it is not even replacing its population. This decline has been attributed to three phenomena. First, there is a high rate of maternal mortality due to relative lack of access to care, the lack of well-trained physicians and nursing staff, and the relative scarcity of technology that could help to address serious pregnancy- and labor-related complications. Second, there is a very high rate of untreated sexually transmitted disease, particularly chlamydia and gonorrhea, resulting in the unknowing infertility of many young women. Third, birth control and abortion were illegal in Technovy for over three decades. As a result, many young women underwent illegal abortions under unsafe conditions, resulting in serious infection, and many had multiple abortions. It is believed that approximately one-third of the country's women of child-bearing age are sterile as a result of repeated abortion procedures under these unsafe conditions.

The government of Technovy is interested in instituting a campaign to encourage surrogate mothering as a means of increasing the birth rate. The procedure of surrogate mothering is defined, for the purpose of this situation, as the insemination of a woman with the sperm of a man to whom she is not married.

When the baby is born, the woman relinquishes her claim to it in favor of another, usually the man from whom the sperm was obtained. Because of your expertise in the ethics of research, the government of Technovy has prevailed upon you to consult with the principal investigator of a prospective cohort study of the individuals involved in surrogate mothering arrangements. The study is to better understand the social and health impact of this type of arrangement.

What ethical issues specific to <u>informed consent</u> are relevant to this study and how can they be resolved? Address this question from (a) the perspective of feminist ethics and (b) a principlistic orientation.

Designing the Informed Consent Process

The Procedure

In many studies, the informed consent process is contemporaneous with actual enrollment: the individual is presented with the information at the same time that he or she is recruited, and, if the individual agrees, he or she is enrolled into the study. Depending upon the nature of the study, the enrollment itself may be a multi-stage process. For instance, eligibility itself may depend not only on an individual's willingness and various demographic characteristics that constitute inclusion and exclusion criteria, such as age, but also on the clinical or laboratory conformation of the presence or absence of a specific disease or condition. In such cases, the investigator may seek preliminary consent from the prospective participant to conduct such confirmatory testing and, following receipt of the results and notification of the participant, seek the individual's consent to further participation, if found to be eligible.

The DCCT Research Group (1989) reported the successful development and utilization of a multi-component program to educate prospective volunteers and enable them to make an informed decision about participation in a complex, long-term clinical trial related to the control of diabetes. Prospective participants recruited through advertising and their physicians underwent an initial eligibility screening to determine if they met age, disease duration, and treatment criteria. Concurrent with the subsequent medical eligibility screening, participants underwent a multi-component informational program, consisting of a slide-tape presentation that described the study and expectations of the participants. Various topics were covered, including the concept of randomization, the standard and experimental study group procedures, eligibility criteria, and risks and benefits. Volunteers were then provided with a booklet covering the material presented in the video in greater depth, and a second booklet that explained relevant scientific terms and procedures in lay language. Both documents were written at the sixth grade reading level. Volunteers were also provided with a consent form to permit continued screening if they wished to participate. Those who did were assigned various behavioral tasks. Upon completion of the medical screening and prior to the presentation of the final informed consent form for randomization, each individual was asked to complete a self-administered test measuring knowledge of the study and procedures and risks and benefits. If the individual scored less than 100 percent, they were provided with additional education in the deficient areas and were then retested. If, on the retest, the individual scored less than 100 percent again, enrollment was at the discretion of the principal investigator. To measure retention of the information, the knowledge test was re-administered at the first annual follow-up visit.

In most clinical trials and intervention studies, the usual approach is to obtain the consent of the prospective participant to his or her participation at the outset. The process of obtaining that consent includes an explanation of randomization and, after giving his or her consent, the individual is then randomized to either the experimental arm or the comparison arm of the study. Zelen (1979), however, proposed an alternative to this process, which is sometimes known as randomized consent. According to this scheme, persons eligible to participate, for instance, in a clinical trial, are randomized before their consent to participate is obtained. Those who are randomized to the standard (comparison) treatment are provided with that treatment without advising them of the alternative, experimental treatment being tested. Those who are randomized to the experimental treatment are provided with an explanation of both treatments. If they should refuse to receive the experimental treatment, they are assigned to the comparison treatment.

Advocates of the Zelen procedure argue that it spares prospective participants the anxiety that accompanies discussions regarding alternative therapies and participation in research. However, this procedure raises both scientific and ethical concerns. From a scientific perspective, it is not possible to conduct the "gold standard" randomized, double-blind trial in which the investigator does not know who is assigned to which treatment group. From an ethical standpoint, only the individuals who are assigned to the experimental arm of the study are actually permitted to make a decision regarding their participation and the receipt of the experimental treatment.

Continuing Consent

The *International Guidelines for the Ethical Review of Biomedical Research Involving Human Subjects* (CIOMS, 1991: 18) provides in commentary to Guideline 3 that:

> The initial consent should be renewed when material changes occur in the conditions or the procedures if the research. For example, new information may come to light, either from inside the study or from outside the study, about the risks or benefits of therapies being tested or about alternatives to the therapies. Subjects should be given such information. In many clinical trials, data are not disclosed to subjects and investigators until the study is concluded. This is ethically acceptable if the data are monitored by a committee responsible for data and safety monitoring…and an ethical review committee has approved their non-disclosure.

Federal regulations provide that, as part of the informed consent process, the investigator may advise participants that they will receive information relating to new developments that may become available during the course of the study that may affect their willingness to participate (45 Code of Federal Regulations section 46.116, 1999).

Respect for persons would seem to require that participants be notified of such developments that could affect their willingness to continue, even if that information relates to a different study. It can be argued that individuals agreed to participate based upon the information that was provided to them, which reflected a specific state of knowledge at the time. When that state of knowledge has shifted, in ways that may be relevant to the participants, whether or not it is relevant to the goal of such research, that change should be made known to avoid what might otherwise be the use of participants as a means only. From a utilitarian perspective, one might assert that the greatest good is to refrain from disclosure because disclosure could result in excessive withdrawals from the study, thereby resulting in termination. Conversely, the greatest good may, indeed, be associated with disclosure because it will reflect and promote integrity in science and open discussion. A casuistic analysis might look to the Tuskegee experiment and the physical and societal harms that ensued from the investigators' failure to advise the subjects of the study of the availability of penicillin for the treatment of syphilis.

Advance Consent

Individuals may wish to indicate now whether they would be willing to participate in research in the future. They may choose to make a present determination relating to a future possibility because there exists the possibility, for each of us, that our mental capacity at that future time will have decreased or become compromised to the extent that we no longer possess the capacity to make that decision. The *International Ethical Guidelines for Biomedical Research Involving Human Subjects* (CIOMS and WHO, 1993: 23) recognizes this possibility and offers a mechanism to address it:

> When it can be reasonably predicted that a competent person will lose the capacity to make valid decisions about medical care, such as in the case of early manifestations of cognitive impairment due to HIV infection or Alzheimer's disease, such a person may be asked to designate the conditions, if any, in which he or she would consent to becoming a research subject while unable to communicate, and to designate a person who will consent on his or her behalf in accordance with the subject's previously expressed wishes.

One mechanism that is available to address this concern in Canada and potentially some jurisdictions in the United States is the advanced directive for research. This is analogous to advance directives in the clinical context, such as a living will or a durable power of attorney for health care. A living will states an individual's wishes with respect to the provision of various types of care that could be potentially available in the future if, at the time that the decision must be made, the individual no longer has the capacity to do so. A durable power of attorney for health care permits an individual to appoint another as his or her agent, to make his or her health care decisions, if the situation should arise in which he or she is unable to do so (Loue, 1995). Depending upon the specific jurisdiction, the appointed individual may be charged with the responsibility of deciding upon a course of action as the now-incapacitated person would have decided (substituted judgment), or may be empowered to decide upon the course of action that would be best for the now-incapacitated individual (best interests). The establishment of such a mechanism has received strong support (Moreno, Caplan, Wolpe, and the Members of the Project on Informed Consent, 1998; Sachs, Rhymes, and Cassel, 1993).

Numerous difficulties attend the interpretation of reliance upon an advance directive for research. For instance, in signifying his or her willingness to participate in "research," did the individual who executed the document intend to limit his or her participation to studies which relate to a disease or condition from which he or she was suffering or which involve a limited level of discomfort or risk? If the agent appointed pursuant to this document does not have an intimate knowledge of the values of the now-incapacitated person, should he or she be permitted to represent the wishes of the individual? If the agent is to decide upon a course of action utilizing a "best interests" standard, how is that standard to be formulated: by whether the individual's condition will be improved or whether it will be substantially improved; by whether the individual's condition will improve without treatment; by whether the risks outweigh the possible benefits of participation (and how will the risks and benefits be defined and prioritized in value); or by which course of action is the least intrusive (Moorhouse and Weisstub, 1996)?

Various arguments have been advanced in support of advance research directives, including the recognition of individual autonomy pursuant to the Nuremberg Code and the larger good that will result from participation in health research. Reliance on a benefit to others as the basis for participation may maximize good, a utilitarian goal, but it may also contravene the Kantian maxim to not use others as a means. Moorhouse and Weisstub (1996) have asserted, as well, that individuals participating in research as a result of having had the foresight to execute such a document will benefit psychologically from knowing that their participation may help others, even if it does not help themselves. However, this assertion is fatuous in that once incapacitated, the individual may have little or no ability to understand what is happening and, consequently, cannot feel altruistic about his or her participation.

Opponents of advance research directives rely on arguments similar to those offered with respect to advance directives in the clinical setting. First, it is contended, the person executing the advance directive is not the same individual who will be affected by the actions embarked upon as a result of that document. In essence, a distinction is made between the once-competent "then-self" and the now-incapacitated "now-self." To impose previously made decisions on what is, in essence, a different person, is to violate the autonomy of the presently-existing individual (Dresser, 1992). Additionally, participation in research must be voluntary and informed. An individual cannot know what the risks and benefits of a not-yet-designed protocol may be, in order to provide informed consent.

Additionally, an individual who lacks capacity may be unable to indicate when he or she wishes to terminate participation in the research study. This situation has been analogized to that of a "Ulysses contract," whereby once participation has begun, it cannot be halted (Macklin, 1987). (The term derives from the situation in which Ulysses wanted to hear the songs of the sirens without jeopardizing his life. He instructed his crew to tie him to the mast and not to release him regardless of his order or the extent of his suffering.) Various suggestions have been made to address this dilemma, including the appointment of a research monitor to safeguard the individual's interests (Moorhouse and Weisstub, 1996), the utilization of a lower level of capacity to revoke consent, and an enhanced role for the review committee in reviewing the consent process and participant enrollment. Moorhouse and Weisstub (1996: 126) have commented on the role of the substitute decision maker (the agent) in this regard:

> An SDM is not merely the person who gives or refuses consent. The SDM as a moral agent must reflect on the nature of the research and consequences of the research participation and must monitor the research to know whether the person should be withdrawn from the study. Ideally, in the research directive there would be instructions about when and why to be withdrawn from a study. Whether withdrawal instructions are prepared in advance or not, the author of the research directive has transferred the authority to withdraw the subject from the study when the subject is at serious risk. Thus, the promise to respect the directive can be overruled by the obligation to protect the person from serious risk of harm by either withdrawing the subject from a study or not permitting entry in the first place if the person's welfare will be at serious risk.

Exercise

You wish to conduct a study to examine numerous factors that may be predictive of the development of Alzheimer's disease. You wish to enroll as participants in the study individuals who have been recently diagnosed with the disease. Study procedures include periodic blood tests, interviews, and neuropsychological assessments, as well as a brain biopsy following the participant's death.

1. What issues arise with respect to the informed consent process specifically due to the nature of the disease that you are studying and the characteristics and potential characteristics of your prospective participants? How will you address each of these issues?
2. Assume for the purpose of this question that, during the course of your study, another study finds a treatment that appears to be effective in ameliorating some of the symptoms of Alzheimer's disease, even in its late stages. Assume further that a large proportion of your participants now lack decision making capacity.
   a. What are the implications of this new finding for your study, if any?

b. What is your obligation to inform your participants of this new development? Refer to ethical theories and principles in your response.
c. Assume that you have decided that you will inform your participants of this finding. How will you do so in view of their impaired cognitive ability?

CONFIDENTIALITY AND PRIVACY CONCERNS

Concern for a participant's privacy and for the confidentiality of the information that he or she provides in the context of research arises regardless of one's ethical framework. A respect for persons would dictate that participants be afforded sufficient privacy and confidentiality to safeguard their interests and ensure that they are not simply viewed as a means. Nonmaleficence suggests that access to information about participants should be restricted so as to minimize the harm that could ensue if it were disclosed. Virtue ethics urges the provision of confidentiality and privacy as the practical manifestation of reasoning with respect to the research situation and feeling for those participating.

The concern for confidentiality and privacy is reflected across the relevant international documents. The Helsinki Declaration provides,

> The right of the research subject to safeguard his or her integrity must always be respected. Every precaution should be take to respect the privacy of the subject and to minimize the impact of the study on the subject's physical and mental integrity and on the personality of the subject.

We will see in chapter 5, however, that there are limits to an investigator's ability to assure confidentiality. The existence of these limitations are recognized by the CIOMS documents, which caution the investigator to provide the research participants with information relating to these restrictions.

> The investigator must establish secure safeguards of the confidentiality of research data. Subjects should be told of the limits to the investigators' ability to safeguard confidentiality and of the anticipated consequences of breaches of confidentiality. (*International Ethical Guidelines for the Review of Biomedical Research Involving Human Subjects*, CIOMS and WHO, 1993: Guideline 12, 35).

> Research may involve collecting and storing data relating to individuals and groups, and such data, if disclosed to third parties, may cause harm or distress. Consequently, investigators should make arrangements for protecting the confidentiality of such data by, for example, omitting information that might lead to the identification of individual subjects, or limiting access to the data, or by other means. It is customary in epidemiology to aggregate

> numbers so that individual identities are obscured. Where group confidentiality cannot be maintained or is violated, the investigators should take steps to maintain or restore a group's good name and status. . . . Epidemiologists discard personal identifying information when consolidating data for purposes of statistical analysis. Identifiable personal data will not be used when a study can be done without personal identification....
> (*International Guidelines for Ethical Review of Epidemiological Research*, CIOMS, 1991: Guideline 26, 17-18).

Various practical mechanisms can be instituted in the context of a research study that will help to reduce the possibility that confidentiality of one or more of the research participants will be breached. This can include the use of unique identifiers. This strategy has been used successfully in California, using Soundex codes in conjunction with AIDS-related research. The Soundex code is composed of a letter which is the first letter of the individual's last name, plus four digits derived from the remaining letters in the person's last name. This code has been called a "numerical alias" (Garfinkel, 1988).

Additional mechanisms include restricting employee and volunteer access to files and to lists containing both the unique identifiers and the individuals' names, using locked file cabinets and locked offices, using passwords or codes to access data stored on computers (Torres, Turner, Harkess, and Istre, 1991), providing employees with comprehensive training on the ethical and legal principles underlying the confidentiality protections and the mechanisms in place in the research project to effectuate these protections (McCarthy and Porter, 1991), and developing an internal procedure for the release of data (Koska, 1992).

MONITORING THE STUDY

The Data Safety Monitoring Board

Data Safety Monitoring Boards (DSMBs) are frequently used in conjunction with clinical trials, and especially those that are double-blinded. The DSMB is an independent committee, usually established in connection with a specified research project, that is charged with the responsibility of monitoring the progress of the trial to protect the safety of the participants and the integrity of the study. To protect participants' safety, the members must be familiar with the protocol, propose appropriate analyses, and periodically review the outcome and safety data. The DSMB helps to maintain the study's integrity through a review of data relating to participant enrollment, quality, and losses to follow-up. Additionally, the DSMB may review study procedures and conduct site visits. Based on such review, the DSMB may suggest revisions in the protocol and/or opereation of the study (National Institutes of Health, 1999). The DSMB's monitoring responsibility is significantly greater than that of an IRB. The DSMB must also determine whether information from the trial should be disseminated to the participants or whether the trial should be terminated due to safety concerns. The DSMB must report its findings on a regular basis to the IRB.

Community Consultation

Community consultation was first introduced into discussions of research ethics in the 1970s (Levine, 1986) due to concerns that individuals who are even mildly vulnerable were at a serious disadvantage in dealing with investigators during the informed consent process, sometimes due to fear of risking a relationship of dependency by questioning the individual who seemingly had greater wisdom and knowledge (Melton, Levine, Koocher, Rosenthal, and Thompson, 1988). Because vulnerability appeared to be enhanced in situations in which each prospective participant was approached individually, it was reasoned that the degree of vulnerability could be reduced by having meetings with numerous prospective participants, at which time they could voice their questions and concerns of the investigators. Such meetings permit the negotiation of those aspects of the study that are amenable to change and an explanation as to why other elements may not be modifiable. The meeting will ultimately inform investigators if the proposed research will be accepted by most members of the community (Melton, Levine, Koocher, Rosenthal, and Thompson, 1988).

Community consultation may also occur through the development of an advisory board to the research project. The board can consist of professional and lay representatives from the community of interest. As an example, Loue, Lloyd, and Phoombour (1996) worked with a community-based advisory board comprised of representatives from various Asian and Pacific Islander communities to develop and sustain a research study and prevention program to reduce HIV risk.

ISSUES RELATING TO DISCLOSURE

Authorship

Morally Tainted Experiments

Various issues relating to authorship may arise. First, there is the question of whether research conducted in an unethical or questionably unethical manner should be published at all. Various international documents provide some guidance in this regard. The Helsinki Declaration states that

> [i]n publication of the results of his or her research, the physician is obliged to preserve the accuracy of the results. Reports of experimentation not in accordance with the principles laid down in this Declaration should not be accepted for publication.

Despite this prohibition, the Uniform Requirements developed by the International Committee of Medical Journal Editors do not require that only research consistent with the Helsinki Declaration be published, or that the researcher must have followed specified ethical rules. It appears, rather, that journals may not require that the authors state in their published articles that the protocol was either consistent with the principles enunciated by the Helsinki

Declaration or that it was approved by an ethical review committee. For instance, Rennie and Yank (1997) examined a total of 53 articles in 5 journals carrying articles on general medicine. Of these 53 articles, 28 (53%) reported informed consent, 22 (42%) reported IRB approval, and 17 (32%) reported that both had been obtained. In 6 studies deemed to have a compelling reason for ethical review, such as those involving repeated testing or the disclosure of genetic information, there was no indication of either informed consent having been obtained or of ethical review by an independent committee. Rikkert and colleagues (1996) found that of 586 interventional studies published in 4 geriatrics journals, only 54 percent reported informed consent and only 40 percent indicated that the study had received the approval of an IRB.

Further, opinion regarding the use and publication of "morally tainted experiments" (Luna, 1997) is divided. Pozos (1992), specifically referring to the Nazi hypothermia experiments, has enumerated the various ethical arguments for and against the publication of data resulting from morally tainted experiments. Arguments in favor include:

1. It is best to get some good from the tainted experiments. Use of the data will advance knowledge.
2. The information gathered is independent of the ethics of the methods; data are not inherently either good or evil.
3. There is no relationship between the characteristics of a researcher and the validity of his or her data.
4. Referencing of the data does not constitute *per se* racism or genocide.

Arguments against such use include:

1. The data are morally tainted; nothing good can come from evil.
2. Reference to the data will acknowledge Nazi philosophy and encourage and support other scientists to conduct similar studies.
3. Referencing the data acknowledges an unethical researcher for contributions to a specified field of endeavor.
4. Referencing the data casts a shadow over scientific research in general.
5. Reference to the data condones the selective dehumanization and death of those who are marginalized within a society.

Freedman (1992: 150) has decried the non-use of the Nazi data as a means of making a statement about the underlying unethical conduct:

> [T]o make a statement, you make a statement, you don't fail to make a statement. Silence is ambiguous and often amounts to the uncomfortably averted gaze, whether intended or not. Statements about the Nazi evil need to be made, in detail, repeatedly, and explicitly, rather than by the indirection of omitting use of the data.

Given these arguments for and against the use of morally tainted data, Greene (1992) has identified three courses of action that are possible: (1) an absolute ban on access to and the use of such data; (2) a laissez-faire approach, by which any investigator who can find the data can use it, subject to whatever restrictions will be imposed by publishers; and 3) selective suppression, whereby guidelines are developed and screening committees established to decide who gets access to what and on what basis. Greene (1992: 1769-170) ultimately concluded:

> [W]e must put the Holocaust and the Nazi experiments directly under the floodlights and on center stage even if some of us and our past and present are partly illuminated by the glare. Instead of banning the Nazi data or assigning it to some archivist or custodial committee, I maintain that it be exhumed, printed, and disseminated to every medical school in the world along with the details of methodology and the names of the doctors who did it, whether or not they were indicted, acquitted, or hanged. It should be taught in the medical school classrooms, not during a special course in ethics or history, but as part of the anatomy, physiology, pathology, microbiology, and pharmacology portion of the curriculum. The data should be presented regularly during grand rounds and research symposia. Let the students and the residents and the young doctors know that this was not ancient history or an episode form a horror movie....It was real. It happened yesterday. It was "medical;" it was "scientific;" it was contemporary with the development of penicillin!

Accordingly, Luna has enumerated three possible courses of action open to an editor who has before him or her a manuscript that describes unethical research: to publish the research, to publish the research with a condemnation of its methods, and to reject the article. Luna has argued for a more mediated approach. In cases of research that reflects serious ethical problems, such as failure to provide participants with complete information or the use of deceit, the manuscript should be rejected. However, in situations in which there are suspicions of ethical problems, the manuscript should be published, together with a discussion of the ethical issues and a rebuttal by the authors of the article. This is in contrast to the view of Angell (1990: 1463) who, as editor of the *New England Journal of Medicine*, stated

> The Journal has taken the position that it will not publish reports of unethical research, regardless of their scientific merit. Only if the work was properly conducted, with full attention to the rights of the human subjects, are we willing to consider it further . . . the approval of the Institutional Review Board (when there is one) and the informed consent of the research subjects are necessary but not sufficient conditions. Even consenting subjects must not be exposed to appreciable risks without the possibility of commensurate benefits.

Participant Confidentiality and Privacy

A second issue relates to participant privacy and confidentiality. The International Committee of Medical Journal Editors (ICMJE) (1995: 311) has published a statement regarding patients' rights to privacy which, by its contents, appears to apply as well to research:

> Patients have a right to privacy that should not be infringed without informed consent. Identifying information should not be published in written descriptions, photographs and pedigrees unless the information is essential for scientific purposes and the patient (or parent or guardian) gives written informed consent for publication. Informed consent for this purpose requires that the patient be shown the manuscript to be published.
>
> Identifying details should be omitted if they are not essential, but patient data should never be altered or falsified in an attempt to attain anonymity. Complete anonymity is difficult to achieve, and informed consent should be obtained if there is any doubt. For example, masking the eye region in photographs of patients is inadequate protection of anonymity.
>
> The requirement for informed consent should be included in the journals' instructions for authors. When informed consent has been obtained, it should be indicated in the published article.

This policy was adopted as the result of complaints from patients who identified themselves in published reports (Colvin-Rhodes, Jellinek, and Macklin, 1978; Murray and Pagan, 1984) changing social expectations, and legal consideration (Smith, 1995). (See chapter 5 for a discussion of the legal implications of these disclosures.)

Snider (1997) has raised several concerns with respect to this policy. First, he notes that the interests of the individual in maintaining privacy and the interest of the public in having additional information are constantly in a state of tension. In situations in which the "patient" is, in fact, the community, such as in a situation involving an outbreak of tuberculosis, must the report remain unpublished because of the index case's unwillingness or inability to give informed consent to publication? Snider also suggest that there may be circumstances in which informed consent for publication should be waived, such as when a manuscript does not contain any identifying information and the details of a case may already be public as a result of media reports. Additional questions that he poses remain unanswered: whether consent should be waived if the provision of the information to the patient may result in harm to the patient, how and when informed consent to publish should be obtained, whether patients who read the manuscript have the right to change or approve it, how disputes between authors and patients will be resolved, and the implications of one patient or relative in a report withholding consent while others provide it.

Doyal (1997) has argued that journals should publish research findings which are based on studies in which participant informed consent was not obtained in three situations: (1) the participants were incompetent to give informed consent, (2) the study utilized medical records only, and (3) the study used human tissue which was the byproduct of surgical intervention. He has placed parameters around each of these situations.

With respect to publication of studies utilizing patients who were incompetent to give informed consent, he argues that such studies should be published if (1) important benefits result from that research, (2) the research could not be completed with healthy volunteers who were able to give informed consent, (3) the risks involved in the therapeutic research were minimal in relation to the risks attending standard therapy, (4) informed consent for children was obtained from someone with parental authority, (5) informed "assent" for incompetent adults was sought from an advocate who was provided with the same information that a participant would have been if he or she were competent, (6) "assent" is not required for therapeutic research with adults where the research involves minimal risk in comparison with the standard treatment, and (7) the purpose and methods of the research are explained to the participant after regaining competence, if that should happen.

Doyal (1997) is willing to permit the publication of studies utilizing medical records without participant consent if access to the records was essential, consent was not practicable, the research is of "sufficient" merit, identifiers have been removed to the extent possible and patients will not be identifiable through publication, contact with the patients is unlikely, permission is obtained from the physician responsible for the patient's care, and researchers who are not clinicians receive instructions regarding their obligation to preserve confidentiality. However, Doyal (1997: 1110) recognizes that this approach is problematic:

> Normally, patients should give their explicit consent for their records to be accessed for this purpose; they should have received appropriate information about who will use them and why and about how confidentiality will be maintained. Yet suppose that the research is epidemiological, that patients might benefit from it in the long term but that for practical reasons informed consent cannot be obtained. Also assume that no further consequences should follow for such patients . . .
>
> *The moral balance here is a fine one. If such research proceeds, there is little doubt that through not obtaining consent a moral wrong is being done. The issue is the degree of this wrong in light of the potential benefit which can follow for the patient-provided that confidentiality is maintained and no further active involvement is expected. Clearly, the public interest will also be served.* (Empahsis added.)

Whether such an approach can be justified within a deontological perspective is questionable, as it would seem here that the individual is being treated but as a means to a specific end. However, it could be argued from a utilitarian perspective that this course of action serves to maximize the greatest good for the greatest number of people.

Accurate Interpretation of the Results

The International Guidelines for Ethical Review of Epidemiological Studies (CIOMS, 1991: paragraph 22, 16) advises:

> Conflict may appear between, on the one hand, doing no harm and, on the other, telling the truth and openly disclosing scientific findings. Harm may be mitigated by interpreting data in a way that protects the interests of those at risk, and is the same time consistent with scientific integrity. Investigators should, where possible, anticipate and avoid misinterpretation that might cause harm.

Qualification as an Author

Most journals require that an individual meet certain minimum standards of participation in the preparation of a manuscript to be legitimately included as an author. For instance, the *British Medical Journal* (Instructions to Authors, 1994: 39) requires that individuals designated as authors have made

> substantial contributions to (a) conception and design, or analysis and interpretation of data; and to (b) drafting the article or revising it critically for important intellectual content; and on (c) final approval of the version to be published. Conditions (a), (b), and (c) must all be met.

The International Committee of Journal Editors, which promulgated this standard (1997), specifies that participation in the acquisition of funding or the collection of data or general supervision of the research group are insufficient to justify authorship. The *American Journal of Public Health* (1996) similarly has established standards for authorship. A letter of transmittal must be sent with the manuscript and must contain all authors' signatures. All authors must have contributed to the conception/design and/or analysis/interpretation of the data, the writing of the manuscript, and the final approval of the completed article and each must be willing to assume public responsibility for the article. All possible sources of conflicts of interest must be disclosed, including funding sources.

Despite these standards, one study conducted in 1989 found that approximately one quarter of the authors of 20 papers had not contributed substantially (Shapiro, Wenger, and Shapiro, 1994). In another study involving 92 authors of 12 papers, of the 84 authors who were not listed as first authors, 61 percent fulfilled the criteria for authorship (Goodman, 1994).

The Source of Support

Significant controversy exists with respect to an investigator's obligation to inform the editors of the journal to which he or she is submitting a manuscript and/or the readers of that journal, of any conflict of interest that may exist with respect to the source of funding for the study at issue. The International Committee of Medical Journal Editors has taken the approach that authors should acknowledge the sources of financial and material support and specify the nature of that support. As Horton (1997b) notes, the concern does not relate to funding per se, but rather to any commercial pressures that may have determined how the data were collected, analyzed, and/or interpreted. The Committee emphasized, in particular, the responsibility of the scientist:

> Scientists have an ethical obligation to submit creditable research results for publication, and should expect to do so. As the persons directly responsible for their work, the authors as individuals should not enter into agreements that interfere with their control over the decision to publish. (Horton, 1997b: 1412).

The editors were cognizant of the potential pressures that funding agencies might exert on the investigators and analogized the involvement of funding agencies to other methodological biases:

> Editors should require authors to describe the role of these sources, if any, in study design, collection, analysis, and interpretation of data, and writing of the report. If the supporting source had no such involvement, the authors should so state. Because the biases potentially introduced by the direct involvement of supporting agencies in research are analogous to methodological biases of other types (eg, study design, statistical and psychological factors, etc), the type and degree of involvement of the supporting agency should be described in the methods section. Editors should also require disclosure of whether or not the supporting agency controlled or influenced the decision to submit the final manuscript for publication. (Horton,1997b: 1412)

Particular concern has been voiced with respect to the potential appearance of a financial or other conflict of interest as the result of posting on the internet and the creation of linkages between sites. The editors cautioned:

> The nature of the Internet requires some special considerations within these well established and accepted policies. As a minimum, sites should indicate the names of editors, authors, and contributors and their affiliations, relevant credentials, and relevant conflicts of interest; documentation and attribution of references and sources for all content; information about copyright; disclosure of site ownership; and disclosure of sponsorship, advertising, and commercial funding.

> Linking from one health or medical Internet site to another may be perceived as a recommendation of the quality of the second site. Journals thus should exercise caution in linking to other sites. If links to other sites are posted as a result of financial considerations, such should be clearly indicated. . . In electronic, as in print layout, advertising and promotional messages should not be juxtaposed with editorial content. Any commercial content should be identified as such. (Horton, 1997b: 1412).

Unanimity does not, however, exist with respect to the likelihood that such disclosure will have its intended effect: the elimination of bias. Horton (1997a) has offered several reasons for this inability: the existing fallacy of objectivity in science, which presupposes the ability of scientists to somehow view their work without bringing to it experiences and biases that they may have; the assumption that financial conflicts of interest are those which are most likely to cause concern; and that disclosure heals any wounds caused by the conflict. Further, Horton suggests that authors may be unwilling to disclose conflicting commitments because of the use of the disparaging term "conflict of interest;" instead, he suggests, the term "dual commitment" should be used.

Several journals, including the *American Journal of Respiratory and Critical Care Medicine*, the *American Journal of Respiratory Cell and Molecular Biology*, and the *Journal of Health Psychology*, have imposed bans on the publication of research sponsored by the tobacco companies (Mark, 1996; Roberts and Smith, 1996). Various authors have criticized this ban as being violative of freedom of the press. Although it might seem that this issue would be more appropriately addressed in the context of a conflict of interest, it raises significant ethical questions with respect to authorship and publication policies.

These questions are reflected in the opposing positions of Caplan (1995) and Englehardt (1995). Caplan (1995: 273-274), arguing in favor of the ban on such publications, has remonstrated:

> Any organization committed to the goal of preventing respiratory illness and disability and to working with government agencies who seek to do everything in their power to reduce the use of tobacco products among children and adults cannot remain credible if it permits research sponsored by the tobacco industry in its publications.

Englehardt (1995: 271-272) has relied on the proverbial "slippery slope" to contest the ban:

> If receiving tobacco money is unacceptable, why is it acceptable to take government money if acquired through unjust taxation policies? Or if the party in power does not support health reform in accord with [the organizations working to prevent respiratory disease]?... Or if the funds come from corporations that sell tobacco products?

Significant pressure may be exerted on scientists either to not accept tobacco industry funds to support research for which other sources of support may be unavailable, or may inadvertently encourage scientists not to disclose their sources of funding where such disclosure is not explicitly required.

Integrity in the Review Process

Riis (1995) has observed the tremendous power that the referee, or reviewer, of an article may wield. As he notes, the review process by its very nature permits a scientist access to the intellectual property of a potential competitor, thereby providing a potential opportunity to dishonestly benefit from another's ideas and work. Unfortunately, some investigators have availed themselves of just such an opportunity. One such instance is described below.

In 1983, Philip Auron's research team cloned and sequenced the expressed gene for a human immune system molecule known as interleukin-1. That research was funded by Cistron Biotechnology, based in Pine Brook, New Jersey. They subsequently submitted a paper to the journal *Nature*, which described their findings. One of the reviewers of that paper was on the staff of Immunex Corp. of Seattle, a key competitor of Cistron. Cistron filed a lawsuit against Immunex, claiming that the reviewer took data from that paper and shared it with his colleagues, who later used it in their own research and a patent application (Marshall, 1995). Cistron and Immunex strongly disagreed about the impropriety of Immunex' reliance on data in the unpublished manuscript if, in fact, it did occur, which Immunex denied. Cistron's representative has argued that, "It would be shockingly unethical and dishonest that a reviewer...should take advantage of a colleague by using the information in a manuscript to his or her own advantage" (Fischer, cited in Taubes, 1996: 1163). Immunex, however, was dismissive of this assertion, claiming that the "[u]se of data in a manuscript to facilitate further research is a practice followed by many scientists....A substantial number of scientists would have made use of the knowledge they obtained from reviewing the Auron manuscript" (Siskind, cited in Taubes, 196: 1163).

The 12-year legal battle ultimately ended with a payment by Immunex to Cistron of $21 million and patent rights to the interleukin-1 protein that was the basis of this dispute. One commentator remarked on the fragility of the peer review process:

> More often than not, there are no faulty data with which to trace misbehaviour, and it is only the decency and honesty of individual researchers that hold the line, and maintain the integrity of science. In particular, much too often, researchers insist on tight confidentiality when they are authors, but are markedly less rigorous when they are sent papers to review. They too need to take note of the dangers. (Anonymous, 1996).

Exercise

Diethylstilbesterol (DES) is a synthetic estrogen. It was first produced in 1938 in London and was used from 1945 through 1971 to prevent spontaneous abortions. Studies of the use of DES in pregnant women were conducted at Harvard University in the late 1940s. The physician-researchers conducting the studies concluded that the use of DES prevented a number of pregnancy complications, although there were no control groups used in these studies. The FDA approved the drug in 1947 for the purpose of preventing miscarriages.

Controlled studies conducted at Tulane University in the 1950s found that more DES-treated women had miscarriages and premature births as compared with the controls. At the University of Chicago's Lying-in Hospital, one half of the pregnant women presenting there received DES and the other half received placebos. They were not told that they were part of a study and they were not told about the drug that they were being given. Twice as many of the DES-using mothers had miscarriages and low birthweight babies compared to those on placebo. The drug continued to be approved and used for another 20 years, despite growing evidence that it was ineffective in preventing miscarriages.

In 1971, an article appeared in the *New England Journal of Medicine*, which reported an association between DES ingestion during pregnancy and the development of clear cell adenocarcinoma in the daughters of the pregnant women who had ingested DES. The FDA banned the use of DES that same year, although by that time, 1.5 million babies had been exposed to it. Subsequent research has found associations between DES ingestion in pregnant women and the following injuries to female offspring: vaginal and cervical dysplasia, adenosis, uterine structural abnormalities, infertility, menstrual irregularities, cervical ridges and cervical erosion, fetal death and premature birth, and breast and reproductive-tract cancers. Injuries to male offspring include genital and semen abnormalities, including penile bleeding, testicular masses, epididymal cysts, hypoplastic testes, and undescended testicles.

1. What, if any, ethical problems exist with respect to the conduct of the DES studies? Be sure to refer to specific provisions of any relevant international documents relating to human experimentation. Explain WHY these provisions, if any, are relevant.

2. Assume for the purposes of this subpart only that you are the editor of a scientific journal to which the Chicago DES research has been submitted for review in 1955. Explain what scientific and ethical concerns might guide your decision regarding publication. In formulating your response, refer to all specific provisions of international documents that may be relevant. Will you publish or not publish this research and why?

3. Assume for the purposes of this subpart only that you are the editor of a scientific journal to which research similar to the Chicago DES research has been submitted for review in 1997. Explain what scientific and ethical concerns might guide your decision regarding publication. In formulating your response, refer to all specific provisions of international documents

that may be relevant. Will you publish or not publish this research and why?

Obligation to Inform

The Findings

The *International Guidelines for Ethical Review of Epidemiological Studies* provide for the dissemination of the research findings to the participants and to the relevant communities. Guideline 13 states:

> Part of the benefit that communities, groups and individuals may reasonable expect from participating in studies is that they will be told of findings that pertain to their health. When findings could be applied in public health measures to improve community health, they should be communicated to the health authorities. In informing individuals of the findings and their pertinence to health, their level of literacy and comprehension must be considered. Research protocols should include provision for communicating such information to communities and individuals.
> Research findings and advice to communities should be publicized by whatever means are available. (CIOMS, 1991: 14)

Further,

> Subjects of epidemiological studies should be advised that it may not be possible to inform them about findings that pertain to their health, but they should not take this to mean that they are free of the disease or condition under study. Often it may not be possible to extract from pooled findings information pertaining to individuals and their families, but when findings indicate a need of health care, those concerned should be advised of means of obtaining personal diagnosis and advice. When epidemiological data are unlinked, a disadvantage to subjects is that individuals at risk cannot be informed of useful findings pertinent to their health. When subjects cannot be advised individually to seek medical attention, the ethical duty to do good can be served by making pertinent health-care advise available to their communities. (CIOMS, 1991: Guideline 14, 14)

Recent "unfortunate events" (Nathan and Weatherall, 1999: 773) called into question the universal acceptance of an investigator's obligation to inform participants of stuffy findings. The events arose at the Hospital for Sick Children in Toronto, where a clinical trial was conducted to "assess the efficacy and safety of the oral chelating agent deferiprone for the prevention of iron overload in patients with transfusion-dependent anemia" (Nathan and Weatherall, 1999: 733). The standard treatment consists of regular blood transfusions together with the administration of chelating agents to prevent death from iron loading. Thalassemias are an increasing health burden for many developing countries (Weatherall and Clegg, 1996).

In 1989, Olivieri at the Hospital for Sick Children commenced a trial to assess the safety and efficacy of deferiprone in transfusion-dependent children with thalassemia. The researchers conducted regular assessment of hepatic iron concentrations to monitor progress. Although short-term data indicated that the drug might control iron accumulation, follow-up indicated that it was poorly controlled in some patients and that the safety threshold was exceeded for increased risk of cardiac disease and early death (Olivieri and Brittenham, 1997).

Apotex Inc., a Canadian pharmaceutical company, had provided funds for the continuation of the study. Despite threats from Apotex to bring legal action if Olivieri disclosed her findings, she presented her findings in refereed journals (Olivieri, 1996; Olivieri, Brittenham, Mclaren, Templeton, Cameron, McClelland, Burt, and Fleming, 1998). Olivieri was subjected to intense pressure, which ultimately resulted in her removal as director of the Toronto Haemoglobinopathies Programme (Nathan and Wetherall, 1999).

Nathan and Weatherall (1999: 772) warned of the lesson to be learned from this:

> [I]n any agreement between a clinical scientist and a company that entails research on patients, the scientist must have the freedom to tell both the patients and, if necessary, the scientific community about any concerns that they have about deleterious effects of the agent or procedure that is being investigated.

Exercise

Assume that you have been asked by the government of Schizovia to collaborate on a phase III clinical trial of Projoy in that country. Schizovia appears to suffer from an extraordinarily high rate of depression in comparison with both developed countries and other underdeveloped countries, although the reasons for this high prevalence are unknown.

1. Assume for the purpose of this question only that, during the course of the clinical trial in Schizovia, another research group discovers a gene that appears to be predictive of the development of clinical depression in a proportion of individuals suffering from the condition. A 1:1 correspondence between having the gene and having depression has not yet been established. Explain the relevance of this research finding, if any, to the conduct and continuation of the clinical trial and any scientific, ethical, and/or legal issues that may arise as a result of this new finding.

2. Assume for the purpose of this subpart only that during the course of the clinical trial, the data that you have been gathering appear to indicate that there is an extremely high rate of intimate partner violence that seems to be associated with depression. However, the study was designed, as you

remember, to test the efficacy of Projoy, not to determine the etiology of the depression. Family violence has not been a subject for open discussion in Schizovia. Explain what ethical obligations you have, if any, to the participants in the study, the government of Schizovia, and the general public of Schizovia with respect to your observation.

Conflict of Interest

Conflicts of interest, financial or otherwise, may arise even during the course of the study. The *International Guidelines for Ethical Review of Epidemiological Studies* (CIOMS, 1991: Guideline 27, 18) cautions investigators that they "should have no undisclosed conflict of interest with their study collaborators, sponsors, or subjects." Apparent conflicts may arise between the investigator and the participant when the investigator believes that the participant should change his or her health behavior, but the participant refuses to do so. These Guidelines indicate that this is not a "true conflict" because the investigators are motivated by the health interests of the participants (CIOMS, 1991: Guideline 30, 18).

There is, however, disagreement as to whether investigators must inform potential participants of a financial conflict of interest. Spiro (1986), for instance, believes that a patients trust in his or her doctor is an important factor in deciding whether or not to participate in a trial, and therefore recommends that the researcher-physician disclose to the patient whether a trial is being funded by a drug company. Jellinek (1982) also advocates this course of action, because it might prompt the patient-prospective participant to ask questions that he or she might not otherwise think to ask about the research. Levine (1982) has maintained that, if disclosure is to be made, it should be made of both private and public sponsors. Finkel (1991) found in a small study of non-patients that disclosure of financial information had little impact on the decision to participate. Rather, the decision making process focused, instead, on the potential risks and benefits associated with participation. Agreement to participate was more likely to be associated with an inability to control the disease and the lack of alternatives.

Exercise

You have been asked to consult with the state of Woeisme in its investigation of the complaints of a particular community in the northern portion of your state. The community believes that it is suffering from an excess number of cancer cases as the result of exposure to toxic substances that have leached into the ground, and subsequently the water supply. In particular, the community is concerned with the large number of leukemia cases among children that has steadily increased over the last 10 to 12 years. It is believed that these substances, which are known carcinogens, have both been dumped illegally by various industrial companies in the area and are also at the federally authorized dumpsites for such substances, which are located nearby. The state epidemiologist, who is responsible for conducting the study, has invested in the stock of a subsidiary company of one of the companies that is alleged to have illegally dumped the carcinogenic substances. The subsidiary is a producer of nutritious foods, such as cereals and high-protein bars. The value of the stock prior to the epidemiologist's assignment to this investigation was approximately $50,000. It is likely that the value of the stock will decrease substantially once the news of the investigation becomes public, and again if the company is ultimately found liable. Discuss any legal and ethical issues that may arise in this situation and how you would resolve them. Justify your response.

Advocacy

The *International Guidelines for Ethical Review of Epidemiological Studies* (1991: 15) provide in Guideline 15:

> Investigators may be unable to compel release of data held by government or commercial agencies, but as health professionals they have an ethical obligation to advocate the release of information that is in the public interest.
>
> Sponsors of studies may press investigators to present their findings in ways that advance special interests, such as to show that a product or procedure is or is not harmful to health. Sponsors must not present interpretations or inferences, or theories or hypotheses, as if they were proven truths.

An advocacy role can take many forms, such as the drafting or support of proposed regulations or legislation or participation in a campaign for a specific change. However, the appropriateness of an advocacy role for researchers is somewhat controversial. Rothman and Poole (1985) have asserted that a researcher's participation in public advocacy is inappropriate. They would entertain the possibility of an epidemiologist participating in the advocacy process only in his or her role as a private citizen (Poole & Rothman, 1990). Last (1991a) holds epidemiologists to the "highest standards of scientific honesty, integrity and impartiality." Although he notes that advocacy, the opposite of impartiality, is often required to address existing health risks, Last resolves the apparent conflict by focusing on the need for sound scientific judgment, rather than impartiality (Last 1991a, 1991b). Weed (1994) justifies advocacy in epidemiology based on the principle of beneficence. Bankowski (1991) explains well the relationship between scientific research and beneficence:

> Epidemiology is a means of quantifying injustice in relation to health care, of monitoring progress towards justice, beneficence, non-maleficence, and respect for persons, as these ethical principles apply to society, and of applying its findings to the control of health problems. That those at the political level charged with safeguarding the public health often neglect or find it inconvenient, or even impractical, to apply epidemiological findings, sometimes because the more vulnerable populations or groups lack the power to assert or safeguard their rights, often

because of the complexity of prioritizing resource allocation, does not invalidate epidemiology. Rather, that this happens is a reason for emphasizing the relation between ethics and human values and health policy-making, and for an ethics of public health, concerned with social justice as well as individual rights, to complement the ethics of medicine.

Gordis (1991) envisions the epidemiologist assuming a societal role in the policy-making process through the presentation of data and its interpretations and the development and evaluation of policy proposals. He acknowledges that there remains a question as to whether a researcher's credibility will be lessened if he or she takes a strong advocacy position on a particular issue. Gordis notes, however, that

> [a]n additional consideration is that since our [epidemiologists'] data have important societal implications, if we [epidemiologists] want society to continue to support our efforts we will have to demonstrate the value of our research for the health of the public. This can only be done if we broaden our responsibility from the research only role to that of policy-related functions. Thus, the epidemiologist must also serve as an educator. Her efforts are directed at many target populations including other scientists, legislators, policy makers, lawyers and judges, and the public. Each must be dealt with differently depending on the specific needs of that population and the objectives towards which the educational effort is directed.

Ultimately, each researcher will have to decide for him- or herself the advisability of participating as an advocate. Cogent arguments support the appropriateness of such a role, but there are clearly professional consequences to be considered in assuming that responsibility.

CHAPTER SUMMARY

This chapter has focused on concerns that may arise during the course of the study, including recruitment issues, ensuring that the prospective participants understand the processes and the risks and benefits, the maintenance of the participants' privacy and the confidentiality of the data, and the safeguarding of participants from risk during the course of the study. Issues relating to participant understanding are complex and necessitate consideration of the individual's ability to understand, to make decisions, and to agree to participate free of duress or coercion. The chapter has addressed in detail issues pertaining to the disclosure of information derived from the study. The theme throughout this chapter has been the protection of individuals who agree to participate in health research.

References

Abt, I. (1903). Spontaneous hemorrhages in newborn children. *Journal of the American Medical Association, 40*, 284-293.

Ackerman, T. (1990). Protectionism and the new research imperative in pediatric AIDS. *IRB, 12*, 1-5.

American College of Physicians. (1989). Cognitively impaired subjects. *Annals of Internal Medicine, 111*, 843-848.

Angell, M. (1990). The Nazi hypothermia experiments and unethical research today. *New England Journal of Medicine, 322*, 1463.

Angrist, B., Peselow, E., Rubenstein, M., Wolkin, A., & Rotrosen, J. (1985). Amphetamine response and relapse risk after depot neuroleptic discontinuation. *Psychopharmacology, 85*, 277-283.

Anonymous. (1996). Peer review and the courts. *Nature, 384*, 1.

Appelbaum, P.S. (1996). Drug-free research in schizophrenia: An overview of the controversy. *IRB: Review of Human Subjects Research, 18*, 1-5.

Ballard, E.L., Nash, F., Raiford, K., & Harrell, L.E. (1993). Recruitment of black elderly for clinical research studies of dementia: The CERAD experience. *Gerontologist, 33*, 561-565.

Bankowski, Z. (1991). Epidemiology, ethics and 'health for all'. *Law, Medicine & Health Care, 19*, 162-163.

Bartholme, W. (1976). Parents, children, and the moral benefits of research. *Hastings Center Report, 6*, 44-45.

Baudouin, J.L. (1990). Biomedical experimentation on the mentally handicapped: Ethical and legal dilemmas. *Medicine and Law, 9*, 1052-1061.

Beecher, H.K. (1970). *Research and the Individual: Human Studies*. Boston: Little, Brown and Co.

Belais, D. (1910). Vivisection animal and human. *Cosmopolitan, 50*, 267-273.

Benson, P.R., Roth, L.H., Appelbaum, P.S., Lidz, C.W., & Winslade, W.J. (1988). Information disclosure, subject understanding, and informed consent in psychiatric research. *Law and Human Behavior, 12*, 455-475.

Bercovici, K. (1921). Orphans as guinea pigs. *The Nation, 112*, 911-913.

Bigorra, J. & Baños, J.E. (1990). Weight of financial reward in the decision by medical students and experienced healthy volunteers to participate in clinical trials. *European Journal of Clinical Pharmacology, 38*, 443-336.

Boffey, P.M. (1987, December 28). Trial of AIDS drug in U.S. lags as too few participants enroll. *New York Times*, p. A1.

Boyd, M.D. and Feldman, R.H.L. Health information seeking and reading and comprehension abilities of cardiac rehabilitation patients. *Journal of Cardiac Rehabilitation*, 4, 343-347.

Brock, D.W. (1994). Ethical issues in exposing children to risks in research. In M.A. Grodin & L.H. Glantz (Eds.). *Children as Research Subjects: Science, Ethics, and Law* (pp. 81-101). New York: Oxford University Press.

Cantwell, A., Jr. (1993). *Queer Blood: The Secret AIDS Genocide Plot*. Los Angeles: Aries Rising Press.

Caplan, A.L. (1995). Con: The smoking lamp should not be lit in ATS/ALA publications. *American Journal of Respiratory and Critical Care Medicine, 151*, 273-274.

Capron, A.M. (1999). Ethical and human-rights issues in research on mental disorders that may affect decision-making capacity. *New England Journal of Medicine, 340*, 1430-1434.

Capron, A.M. (1991). Protection of research subjects: Do special rules apply in epidemiology? *Law, Medicine & Health Care, 19,* 184-190.

Cassileth, B.R., Zupkis, R.V., Sutton-Smith, K., & March, V. (1980). Informed consent—Why are its goals imperfectly realized? *New England Journal of Medicine, 302*, 896-900.

Christakis, N.A. (1996). The distinction between ethical pluralism and ethical relativism: Implications for the conduct of transcultural clinical research. In H.Y. Vanderpool (Ed.). *The Ethics of Research Involving Human Subjects: Facing the 21st Century* (pp. 261-280). Frederick, Maryland: University Publishing Group, Inc.

Christakis, N.A. (1988). The ethical design of an AIDS vaccine trial in Africa. *Hastings Center Report, 18*, 31-37.

Cimons, M. (1989, November 9). Lack of volunteers may hurt AIDS drug trials. *Los Angeles Times*, p. A5.

Colvin-Rhodes, L.M., Jellinek, M., & Macklin, R. (1978). Studying grief without consent. *Hastings Center Report, August*, 21-22.

Committee on Bioethics, Academy of Pediatrics. (1998). Informed consent, parental permission, and assent in pediatric practice. *Journal of Child and Family Nursing, 1*, 57-61.

Council for International Organizations of Medical Sciences (CIOMS), World Health Organization (WHO). (1993). *International Guidelines for Biomedical Research Involving Human Subjects.* Geneva: WHO.

Council for International Organizations of Medical Sciences (CIOMS). (1991). *International Guidelines for Ethical Review of Epidemiological Studies.* Geneva: Author.

Dal-Re, R. (1992). Elements of informed consent in clinical research with drug: A survey of Spanish clinical investigators. *Journal of Internal Medicine, 231*, 375-379.

Davidson, M., Keefe, R.S.E., Mohs, R.C., Siever, L.J., Losonczy, M.F., Horvath, T.B., & Davis, K.L.. (1987). L-Dopa challenge and relapse in schizophrenia. *American Journal of Psychiatry, 144*, 934-938.

Davis, J.M. (1985). Maintenance therapy and the natural course of schizophrenia. *Journal of Clinical Psychiatry, 11*, 18-21.

Davis, J.M. (1975). Overview: Maintenance therapy in psychiatry. I. Schizophrenia. *American Journal of Psychiatry, 132*, 1237-1245.

DCCT Research Group. (1989). Implementation in a multicomponent process to obtain informed consent in the Diabetes Control and Complications Trial. *Controlled Clinical Trials, 10*, 83-96.

Director, Division of Human Subject Protections, OPRR (195, November 9). Memorandum to professional staff, Division of Human Subject Protections, OPRR, Obtaining and Documenting Informed Consent of Subjects Who Do Not Speak English. Available at http://grants.nih.gov/grants/oprr/humansubjects/guidance/ic-non-e.htm

Doak, L.G. and Doak, C.C. (1987). Lowering the silent barriers to compliance for patients with low literacy skills. *Promoting Health, 8*, 6-8.

Doak, L.G. and Doak, C.C. (1980). Patient comprehension profiles: Recent findings and strategies. *Patient Counseling and Health Education, 2*, 101-106.

Doyal, L. Informed consent in medical research: Journals should not publish research to which patients have not given fully informed consent—with three exceptions. *British Medical Journal, 314*, 1107-1111.

Dresser, R.S. (1992). autonomy revisited: The limits of anticipatory choices. In R.H. Binstock, S.G. Post, & P.J. Whitehouse (Eds.). *Dementia and Aging: Ethics, Values and Policy Choices* (pp. 71-85). Baltimore: Johns Hopkins University Press.

El-Sadr, W. & Capps, L. (1992). The challenge of minority recruitment in clinical trials for AIDS. *Journal of the American Medical Association, 267*, 954-957.

Englehardt, H.T. (195). Pro: The search for untainted money. *American Journal of Respiratory and Critical Care Medicine, 151*, 271-272.

Epstein, L.C. & Lasagna, L. (1969). Obtained informed consent. *Archives of Internal Medicine, 123*, 682-688.

Expert Panel Report to the National Institutes of Health. (1998, February). Research Involving Individuals with Questionable Capacity to Consent: Ethical Issues and Practical Considerations for Institutional Review Boards (IRBs).

Finkel, M.J. (1991). Should informed consent include information on how research is funded? *IRB, 13*, 1-3.

Freedman, B. (1992). Moral analysis and the use of Nazi experimental results. In A.L. Caplan (Ed.). *When Medicine Went Mad: Bioethics and the Holocaust* (pp. 141-154). Totowa: New Jersey: Humana Press.

Freedman, B., Fuks, A., & Weijer, C. (1993). In loco parentis: Minimal risk as an ethical threshold for research upon children. *Hastings Center Report, 23*, 13-19.

Garfinkel, S.L. (1988). AIDS and the Soundex code. *IRB, 10*, 8-9.

Gaylin, W. (1982). Competence: No longer all or none. In W. Gaylin and R. Macklin (Eds.). *Who Speaks for the Child: The Problems of Proxy Consent* (pp. 27-54). New York: Plenum Press.

Goodman, N.W. (194). Survey of fulfilment of criteria for authorship in published medical research. *British Medical Journal, 309*, 1482.

Gordis, L. (1991). Ethical and professional issues in the changing practice of epidemiology. *Journal of Clinical Epidemiology, 44,* Supp. 1, 9S-13S.

Gostin, L. (1991). Ethical principles for the conduct of human subject research: Population-based research and ethics. *Law, Medicine & Health Care, 19*, 191-201.

Greene, V.W. (1992). Can scientists use information derived from concentration camps? Ancient answers to new questions. In A.L. Caplan (Ed.). *When Medicine Went Mad: Bioethics and the Holocaust* (pp. 155-170). Totowa: New Jersey: Humana Press.

Guinan, M.E. (1993). Black communities' belief in AIDS as "genocide": A barrier to overcome for HIV prevention. *Annals of Epidemiology, 3*, 193-195.

Hall, A.J. (1989). Public health trials in West Africa: Logistics and ethics. *IRB, 11*, 8-10.

Hammill, S.M., Carpenter, H.C., & Cope, T.A. (1908). A comparison of the von Pirquet, Calmette and Moro tuberculin tests and their diagnostic value. *Archives of Internal Medicine, 2*, 405-447.

Horton, R. (1997a). Conflicts of interest in clinical research: opprobrium or obsession? *Lancet, 349*, 1112-1113.

Horton, R. (1997b). Sponsorship, authorship, and a tale of two media. *Lancet, 349*, 1411-1412.

Instructions to Authors. (1994). *British Medical Journal, 308*: 39-42.

International Committee Of Medical Journal Editors. (1995). Protection of patients' rights to privacy. *British Medical Journal, 311*, 1272.

International Committee of Medical Journal Editors. (1997). Uniform requirements for manuscripts submitted to biomedical journals. *Journal of the American Medical Association, 277*, 927-934.

Janosky, J. & Starfield, B. (1981). Assessment of risk in research in children. *Journal of Pediatrics, 98*, 842-846.

Jellinek, M.S. (1982). IRBs and pharmaceutical company funding. *IRB, 4*, 9-10.

Katz, J. (1993). Human experimentation and human rights. *St. Louis University Law Journal, 38*, 7-54.

Kayser-Boyd, N., Adelman, H., Taylor, L., & Nelson, P. (1986). Children's understanding of risks and benefits of psychotherapy. *Journal of Clinical Child Psychology, 15*, 165-171.

Keyserlingk, E.W., Glass, K., Kogan, S., & Gauthier, S. (1995). Proposed guidelines for the participation of persons with dementia as research subjects. *Perspectives in Biology and Medicine, 38*, 319-361.

Keith-Spiegel, P. & Maas, T. (1981). Consent to research: Are there developmental differences. Proceedings of the American Psychological Association. Cited in S. Leikin. (1993). Minors' assent, consent, and dissent to medical research. *IRB, 15*, 1-7.

Klerman, G.L. (1977). Development of drug therapy for the mentally ill. *Federation Proceedings, 36*, 2352-2355.

Klonoff, E.A. & Landrine, H. (1999). Do blacks believe that HIV/AIDS is a government conspiracy against them? *Preventive Medicine, 28*, 451-457.

Kolata, G. (1988, December 18). Recruiting problems in New York slowing U.S. trial of AIDS drugs. *New York Times*, p. 1-1.

Koska, M.T. (1992, January 5). Outcomes research: Hospitals face confidentiality concerns. *Hospitals*, 32-34.

Lara, M.D.C. & de la Fuente, J.R. (1990). On informed consent. *Bulletin of Pan-American Health Organization, 24*, 419-424.

Last, J. M. (1991a). Epidemiology and ethics. *Law, Medicine & Health Care, 19*, 166-173.

Last, J. M. (1991b). Obligations and responsibilities of epidemiologists to research subjects. *Journal of Clinical Epidemiology, 44*, Supp. 1, 95S- 101S.

Lavelle-Jones, C., Byrne, D. J., Rice, P., & Cuschieri, A. (1993). Factors affecting quality of informed consent. *British Medical Journal, 396*, 885-890.

Leikin, S. (1993). Minors' assent, consent, or dissent to medical research. *IRB, 15*, 1-7.

Levine, C. (1989, November 15). AIDS crisis sparks a quiet revolution. *Los Angeles Times*, p. B7.

Levine, C. (1991). Children in HIV/AIDS clinical trials. Still vulnerable after all these years. *Law, Medicine & Health Care, 19*, 231-237.

Levine, R.J. (1992). Clinical trials and physicians as double agents. *Yale Journal of Biology and Medicine, 65*, 65-74.

Levine, R.J. (1982). Comment. *IRB, 4*, 9-10.

Levine, R.J. (1986). *Ethics and Regulation of Clinical Research* (2nd ed.). Baltimore: Urban and Schwarzenberg.

Lewis, C., Lewis, M., & Ifekwunigue, M. (1978). Informed consent by children and participation in an influenza vaccine trial. *American Journal of Public Health, 68*, 1079-1082.

Lieberman, J.A., Kane, J.M., Sarantakos, S., Gadaletta, D., Woerner, M., Alvir, J., & Ramos-Lorenzi, J. (1986). Influences of pharmacologic and psychosocial factors in the course of schizophrenia: Prediction of relapse in schizophrenia. *Psychopharmacology Bulletin, 22*, 845-853.

Loue, S. (1995). Living wills, durable powers of attorney and HIV infection: The need for statutory reform. *Journal of Legal Medicine, 16*, 461-480.

Loue, S., Lloyd, L.S., & Phoombour, E. (1996). Organizing Asian Pacific Islanders in an urban community to reduce HIV risk: A case study. *AIDS Education and Prevention, 8*, 381-193.

Loue, S., Okello, D., & Kawumu, M. (1996). Research bioethics in the Uganda context: A program summary. *Journal of Law, Medicine & Ethics, 24*, 47-53.

LoVerde, M.E., Prochazka, A.V., & Byyny, R.L. (1989). Research consent forms: Continued unreadability and increasing length. *Journal of General Internal Medicine, 4*, 410-412.

Luna, F. (1997). Vulnerable populations and morally tainted experiments. *Bioethics, 11*, 256-264.

Macklin, A. (1987). Bound to freedom: The Ulysses contract and the psychiatric will. *University of Toronto Faculty Law Review, 45*, 37-68.

Marks, D.F. (1996). A higher principle is at stake than simply freedom of speech. *British Medical Journal, 312*, 773-774.

Marshall, E. (1995). Suit alleges misuse of peer review. *Science, 270*, 1912-1914.

Martin, J.M. & Sacks, H.S. (1990). Do HIV-infected children in foster care have access to clinical trials of new treatments? *AIDS & Public Policy Journal, 5*, 3-8.

McCarthy, C.R. & Porter, J.P. (1991). Confidentiality: The protection of personal data in epidemiological and clinical research trials. *Law, Medicine & Health Care, 19*, 238-241.

Melnick, V.L., Dubler, N., Weisbard, A. & Butler, R. (1985). Clinical research in senile dementia of the Alzheimer's type: Suggested guidelines addressing the ethical and legal issues. In V.L. Melnick & N.N. Dubler (Eds.). *Alzheimer's Dementia* (pp. 295-308). New Jersey: Humana Press.

Melton, G. (1980). Children's concepts of their rights. *Journal of Clinical Child Psychology, 9*, 186-190.

Melton, G.B. (1989). Ethical and legal issues in research and intervention. *Journal of Adolescent Health Care, 10*, 36S-44S.

Melton, G.B., Levine, R.J., Koocher, G.P., Roenthal, R., & Thompposon, W.C. (1988). Community Consultation in Socially Sensitive research: Lessons from clinical trials of treatments for AIDS. *American Psychologist, 43*, 573-581.

Merton, V. (1993). The exclusion of pregnant, pregnable, and once-pregnable people (a.k.a. women) from biomedical research. *American Journal of Law & Medicine, 19*, 369-451.

Mirkin, B.L. (1975). Drug therapy and the developing human: Who cares? *Clinical Research, 23*, 110-111.

Moreno, J., Caplan, A.L., Wolpe, P.R., & Members of the Project on Informed Consent, Human Research Group. (1998). Updating protections for human subjects involved in research. *Journal of the American Medical Association, 280*, 1951-1958.

Morrow, G., Gootnick, J., & Schmale, A. (1978). A simple technique for increasing cancer patients' knowledge of informed consent to treatment. *Cancer, 42*, 793-799.

Murray, J.C. & Pagan, R.A. (1984). Informed consent for research publication of patient-related data. *Clinical Research, 32*, 404-408.

Muss, H.B., White, D.R., Michielutte, R., Richards, F., II, Cooper, M.R., Williams, S., Stuart, J.J. & Spurr, C.L. (1979). Written informed consent in patients with breast cancer. *Cancer, 43*, 1549-1556.

Nathan, D.G. & Weatherall, D.J. (1999). Academia and industry: Lessons from the unfortunate events in Toronto. *Lancet, 353*, 771-772.

National Bioethics Advisory Commission. (1998, November 12). Research involving persons with mental disorders that may affect decision making capacity. Rockville, Maryland: Author.

National Commission for the Protection of Human Subjects of Biomedical and Behavioral Research. (1978). *The Belmont Report: Ethical Principles and Guidelines for the Protection of Human Subjects of Research.* (DHEW Publication No. (OS) 78-0012). Washington, D.C.: Department of Health, Education and Welfare.

National Institutes of Health. (1999, June 11). Guidance on reporting adverse events to institutional review boards for NIH-supported multicenter clinical trials. Available at http://grants.nih.gov/grants/guide/notice-files/not99-107.html.

National Institutes of Health. (no date). Interim—Research Involving Individuals with Questionable Capacity to Consent: Points to Consider. Available at http://grants.nih.gov/grants/policy/questionablecapacity.html.

National Institutes of Health. (1998, March 6), NIH Policy and Guidelines on the Inclusion of Children as Participants in Research Involving Human Subjects. Available at http://grants.nih.gov/grants/guide/notice-files/not98-024.html.

Newton, L. (1990). Ethical imperialism and informed consent. *IRB, 12*, 11.

Olivieri, N.F. (1996). Long-term follow-up of body iron inpatients with thalassemia major during therapy with the orally active iron chelator deferiprone (l1). *Blood, 88*, 310A.

Olivieri, N.F. & Brittenham, G.M. (1997). Iron-chelating therapy and the treatment of thalassemia. *Blood, 89*, 739-761.

Olivieri, N.F., Brittenham, G.M., McLaren CE, Templeton, D.M., Cameron, R.G., McClelland, R.A., Burt, A.D., & Fleming, K.A. (1998). Long-term safety and effectiveness of iron-chelation therapy with deferiprone for thalassemia major. *New England Journal of Medicine, 339*, 417-423.

Ondrusek, N., Abramovitch, R., & Koren, G. (1992, May). Children's ability to volunteer for nontherapeutic research. *Proceedings of the American Pediatric Society and the Society for Pediatric Research.*

Pellegrino, E.D. (1992). Beneficence, scientific autonomy, and self-interest: Ethical dilemmas in clinical research. *Cambridge Quarterly of Healthcare Ethics, 1*, 361-369.

Penslar, R.L. (no date). *Institutional Review Board Guidelines*. Available at http://grants.nih.gov/grants/oprr/irb/irbintroduction.html.

Peterson, B.T., Clancy, S.J., Champion, K., & McLarty, J.W. (1992). Improving readability of consent forms: What the computers may not tell you. *IRB, 14*, 6-8.

Poole, C. & Rothman, K. J. (1990). Epidemiologic science and public health policy (letter). *Journal of Clinical Epidemiology, 43,* 1270.

Pozos, R.S. (1992). Scientific inquiry and ethics: The Dachau data. In A.L. Caplan (Ed.). *When Medicine Went Mad: Bioethics and the Holocaust* (pp. 95-108). Totowa: New Jersey: Humana Press.

Ramsey, R. (1989). Consent as a canon of loyalty with special reference to children in medical investigations. In T.A. Mappes & J.S. Zembaty (Eds.). *Biomedical Ethics* (3rd ed.., pp. 222-224). New York: McGraw-Hill, Inc.

Redmon, R.B. (1986). How children can be respected as 'ends' and yet still be used as subjects in nontherapeutic research. *Journal of Medical Ethics, 12*, 77-82.

Rennie, D. & Yank, V. (1997). Disclosure to the reader of institutional review board approval and informed consent. *Journal of the American Medical Association, 277*, 922-923.

Riis, P. (1995). Ethical issues in medical publishing. *British Journal of Urology, 76*, 1-4.

Rikkert, M., ten Have, H., & Hoefnagels, W. (1996). Informed consent in biomedical studies of aging: Survey of four journals. *British Medical Journal, 313*, 1117-1120.

Roach, W.H., Jr. & Aspen Health Law Center. (1994*). Medical Records and the Law* (2nd ed.). Gaithersburg, Maryland: Aspen.

Roberts, J. & Smith, R. (1996). Publishing research supported by the tobacco industry. *British Medical Journal, 312*, 193-194.

Rothman, K. J. & Poole, C. (1985). Science and policy making. *American Journal of Public Health, 75,* 340-341.

Sachs, G.A., Rhymes, J., & Cassel, C.K. (1993). Biomedical and behavioral research in nursing homes: Guidelines for ethical investigations. *Journal of the American Geriatrics Society, 41*, 771-777.

Schoepf, B.G. (1991). Ethical, methodological, and political issues of AIDS research in Central Africa. *Social Science & Medicine, 33*, 749-763.

Schooler, N.R. & Levine, J. (1983). Strategies for enhancing drug therapy of schizophrenia. *American Journal of Psychotherapy, 37*, 521-532.

Schwartz, A.H. (1972). Children's concepts of research hospitalization. *New England Journal of Medicine, 287,* 589-592.

Shamoo, A.E. & Irving, D.N. (1997). Accountability in research using persons with mental illness. In A.E. Shamoo (Ed.). *Ethics in Neurobiological Research with Human Subjects: The Baltimore Conference on Ethics*. Amsterdam: Overseas Publishers Association.

Shamoo, A.E. & Keay, T.J. (1996). Ethical concerns about relapse studies. *Cambridge Quarterly of Healthcare Ethics, 5*, 373-386.

Shapiro, D.W., Wenger, N.S., Shapiro, M.F. (1994). The contributions of authors to multiauthored biomedical research papers. *Journal of the American Medical Association, 271*, 438-442.

Shavers-Hornaday, V.L., Lynch, C.F., Burmeister, L.F., & Torner, J.C. (1997). Why are African Americans under-represented in medical research studies? Impediments to participation. *Ethnicity and Health, 2*, 31-45.

Silva, M.C. & Sorrell, J.M. (1988). Enhancing comprehension of information for informed consent: A review of empirical research. *IRB, 10*, 1-5.

Silverman, W.A. (1989). The myth of informed consent: In daily practice and in clinical trials. *Journal of Medical Ethics, 15*, 6-11.

Smeltezer, S.C. (1992). Women and AIDS: Sociopolitical issues. *Nursing Outlook, 40*, 152-157.

Snider, D.E. (1997). Patient consent for publication and the health of the public. *Journal of the American Medical Association, 278*, 624-626.

Spiro, H.M. (1986). Mammon and medicine: The reward of clinical trials. *Journal of the American Medical Association, 255*, 1174-1175.

Steinbrook, R. (1989, September 25). AIDS trials shortchange minorities and drug users. *Los Angeles Times*, p. 1-1.

Sugarman, J., Kass, N.E., Goodman, S.N., Perentesis, P., Fernandes, P. & Faden, R.R. (1998). What patients say about medical research. *IRB, 20*, 1-7.

Sunderland, T. & Dukoff, R. (1997). Informed consent with cognitively impaired patients: An NIMH perspective on the durable power of attorney. In A.E. Shamoo (Ed.). *Ethics in Neurobiological Research with Human Subjects: The Baltimore Conference on Ethics* (pp. 229-238). Amsterdam: Overseas Publishers Association.

Susman, E.J., Dorn, L.D., & Fletcher, J.C. (1992). Participation in biomedical research: The consent process as viewed by children, adolescents, young adults, and physicians. *Journal of Pediatrics, 121*, 547-552.

Tankanow, R.M., Sweet, B.V., & Weiskopf, J.A. (192). Patients' perceived understanding of informed consent in investigational drug studies. *American Journal of Hospital Pharmacy, 49*, 633-635.

Taub, H.A. (1986). Comprehension of informed consent for research: Issues and directions for future study. *IRB, 8*, 7-10.

Taubes, G. (1996). Trial set to focus on peer review. *Science, 273*, 1162-1164.

T.D. v. New York Office of Mental Health, Supreme Court of New York, Appellate Division, First Department, December 1996.

Thomas, S.B. & Quinn, S.C. (1991). The Tuskegee syphilis study, 1932 to1972: Implications for HIV education and AIDS risk reduction programs in the black community. *American Journal of Public Health, 81*, 1498-1504.

Thong, Y.H. & Harth, S.C. (1991). The social filter effect of informed consent in clinical research. *Pediatrics, 87*, 568-569.

Tobias, J.S. (1988). Informed consent and controlled trials. *Lancet, 2*, 1194.

Torres, C.G., Turner, M.E., Harkess, J.R., & Istre, G.R. (1991). Security measures for AIDS and HIV. *American Journal of Public Health, 81*, 210-211.

United States Department of Education. (1986*). Digest of Education Statistics 1985-1986.* Washington, D.C.: United States Department of Education Center for Statistics.

Van Kammen, D.P., Docherty, J.P., & Bunney, W.F. Jr. (1982). Prediction of early relapse after pimozide discontinuation by response to α-amphetamine during pimozide treatment. *Biological Psychiatry, 17*, 233-242.

Vollmer, W.M., Hertert, S., & Allison, M.J. (1992). Recruiting children and their families for clinical trials: A case study. *Controlled Clinical Trials, 13*, 315-320.

Weatherall, D.J. & Clegg, J.B. (1996). Thalassemia—a global public health problem. *Nature & Medicine, 2*, 847-849.

Weithorn, L. & Campbell, S. (1982). The competency of children and adolescents to make informed decisions. *Child Development, 53*, 1589-1598.

Welton, A.J., Vickers, M.R., Cooper, J.A., Meade, T.W., & Marteau, T.M. (1999). Is recruitment more difficult with a placebo arm in randomized controlled trials? A quasirandomized interview based study. *British Medical Journal, 318*, 1114-1117.

Wender, E.H. (1994). Assessment of risk to children. . In M.A. Grodin & L.H. Glantz (Eds*.). Children as Research Subjects: Science, Ethics, and Law* (pp. 182-192). New York: Oxford University Press.

What AJPH authors should know. (1996). *American Journal of Public Health*, 86, 1686.

World Health Organization (WHO). (1989, February 27-March 2). Criteria for international testing of candidate HIV vaccines.

Young, D.R., Hooker, D.T., & Freeberg, F.E. (1990). Informed consent documents: Increasing comprehension by reducing reading level. *IRB, 12*, 1-5.

Zelen, M. (1979). A new design for randomized clinical trials. *New England Journal of Medicine, 300*, 1242-1245.

45 Code of Federal Regulations §§ 46.304-.305 (1999).

45 Code of Federal Regulations §§ 46.404-46.416 (1999).

5
LEGAL ISSUES IN RESEARCH

INTRODUCTION

When we speak of legal issues in research, it is critical to recognize not only the distinction between legal and ethical responsibilities, but also the fact that ethical and legal obligations may not be congruent. For instance, the section of this chapter discussing the Freedom of Information Act explains how that legislation was used to acquire information relating to a molecular analysis performed by the Centers for Disease Control in connection with the possible transmission of HIV from a dentist to his patient. Although there was no question that this usage was legal, there was considerable debate about whether the researchers' use of the Freedom of Information Act to obtain the data was ethical. Accordingly, the analysis of a situation to identify and address legal issues is not sufficient; a situation must also be examined to identify and address any potential ethical concerns.

A discussion of the legal consequences of improper conduct in scientific research is somewhat confusing because administrative responsibility for the oversight of the conduct and the corresponding procedures depend upon the nature of that conduct. Conduct that is classifiable as scientific misconduct within the federal definition of the term is within the jurisdiction of the Office of Research Integrity (ORI) of the Health and Human Services Administration. The misuse of human participants and animal subjects in research is addressed by the Office for Protection from Research Risks (OPRR) at the National Institutes of Health. Conduct in connection with regulated research, such as the testing and evaluation of human and animal drugs, food, and food additives and the testing and evaluation of medical devices is under the jurisdiction of the Food and Drug Administration (FDA). Institutions conducting research, however, may or may not distinguish between types of conduct in this manner in investigating alleged instances of misconduct or abuse. And, if the case is followed through to litigation, the court procedures that are relevant depend on the civil or criminal nature of the action, and not the classification as delineated above. For ease of reference, this chapter discusses conduct in scientific research utilizing the administrative classification scheme.

MISUSE OF HUMAN PARTICIPANTS IN RESEARCH

As indicated above, responsibility for the monitoring and evaluation of compliance by institutions with rules governing research involving human subjects lies with the Office of Protection from Research Risks (OPRR). Although that office is currently within the National Institutes of Health, OPRR will be relocated in the Office of the Secretary of the Department of Health and Human Services (United States Department of Health and Human Services, 1999, Nov. 4). This move may reduce the perception of a conflict of interest on the part of OPRR because of its position within NIH and its concomitant obligation to review research conducted by NIH researchers (Office for Protection from Research Risks Review Panel, 1999).

In order to receive HHS support for research involving human participants, recipient institutions must furnish the HHS with written Assurances of Compliance that describe how they will comply with the HHS regulations and the PHS policy governing research. These assurances are negotiated between OPRR and the institution; all such Assurances must be consistent with the regulations and the PHS policy. Although HHS regulations apply only to HHS/PHS-funded research, the negotiation of an Assurance of Compliance obliges the funded institution to apply the same ethical principles in all of its research.

In order to fulfill its monitoring and evaluation responsibilities, OPRR evaluates allegations of noncompliance with the HHS regulations or PHS policy. Institutional officials, committees, researchers, and other institutional agents are expected to comply with the Assurances and, accordingly, with the regulations and policy upon which the Assurances are predicated.

OPRR advises institutional officials of its initiation of an investigation. Generally, evaluations proceed through the following steps:

1. OPRR receives or discovers an allegation of indication of noncompliance. The allegation may come directly from an institution, which is required under the regulations and the policy to report any serious or continuing noncompliance to OPRR.
2. OPRR will determine whether or not it has jurisdiction based on the receipt of HHS support to the institution or on an applicable Assurance of Compliance.
3. OPRR will acknowledge the institution's report of noncompliance, if that was the source of the indication. Alternatively, OPRR will notify the designated official at the institution of the possible noncompliance and will request that the institution investigate the situation and submit a report to OPRR by a specified date. OPRR may also notify the specific investigator if the allegation involved a specified individual.
4. OPRR will evaluate the institution's report and any other information that is available with respect to the matter.
5. Where possible, OPRR will attempt to resolve the noncompliance through correspondence with the institution. In such cases, OPRR will advise the complainant of the ultimate outcome of the investigation.
6. If OPRR decides that a formal report of findings is necessary, OPRR will notify the appropriate institutional official of this decision. Copies of the report are provided to the authorized official of the institution and the complainant, together with a request that errors of fact be indicated.
7. OPRR will attach the institutional or individual identification of errors of fact to its report and will forward the final report to the institution and to the complainant. Errors of fact are addressed in the preface of the report.
8. The final report is available to the public through the Freedom of Information Act (Director, OPRR, 1997).

(See below for a discussion of the Freedom of Information Act.) Records that can be retrieved by an individual's name or other identifier are generally not disclosable pursuant to the provisions of the Privacy Act. (See below.)

In conducting its evaluations, OPRR has noted that the following problems appear to be common: deficiencies in informed consent documents, use of overly complex language in informed consent documents, the inclusion of exculpatory language in informed consent documents, and the reliance on boiler plate informed consent documents. Many of the problems relate to the responsibilities of the institutional review committees: the failure to review protocols that require review, the conduct of business in the absence of a quorum, the failure to maintain diversity in the membership of the review committee, poor maintenance of the records, a lack of appropriate policies and procedures, and the inadequate reporting of problems in research (Compliance Oversight Branch, 1999).

The action ultimately taken by OPRR in a specific case will be designed to remedy the noncompliance and "to foster the best interests of human research subjects . . . the institution, the research community, and the HHS or PHS funding component" (Director, OPRR , 1997: 2). The oversight evaluation may result in one or more of the following actions or determinations:

1. the protections under an institution's Assurance of Compliance are in compliance with the HHS regulations or PHS policy,
2. the protections under an institution's Assurance of Compliance are in compliance with the HHS regulations or PHS policy, but improvements are recommended,
3. the approval of the institution's Assurance of Compliance will be restricted and HHS cannot provide research support until the terms of the restriction, such as a requirement for special reporting to OPRR, have been satisfied,
4. OPRR may withdraw its approval of the institution's Assurance of Compliance,
5. OPRR may recommend to HHS officials or PHS agency heads that a particular investigator be suspended or removed and/or that peer review groups be notified of the investigator's or institution's noncompliance prior to their review of new projects, and/or
6. OPRR may recommend to HHS that institutions or investigators be found ineligible to participate in HHS-funded research, i.e. debarment (Director, OPRR, 1997).

As an example, OPRR recently ordered the suspension of all research at Duke University Medical Center due to concerns centering on the informed consent procedures and the failure to report the occurrence of an unexpected injury to a research volunteer (Hilts and Stolberg, 1999).

The provisions relating to debarment from federal grants and contracts apply to all individuals that have participated, are currently participating, or may reasonably be expected to participate in transactions under federal procurement programs (45 Code of Federal Regulations section 76.110, 1999). Debarment may be instituted as a sanction against an investigator in order to protect the public interest (45 Code of Federal Regulations section 76.115, 1999). An investigator may be debarred for any of the following reasons: (1) conviction of or civil judgment for commission of fraud violation of federal or state antitrust laws, the commission of embezzlement, theft, forgery, bribery, making false statements, receiving stolen property, making false claims, or obstruction of justice, or the commission of any other offense that may indicate a lack of business integrity or honesty that directly affects the person's responsibility, or (2) violation of the terms of a public agreement or transaction that is so serious that it affects the integrity of an agency program (45 Code of Federal Regulations section 76.305, 1999).

Following the reporting of a potential cause for debarment, the agency will issue a notice of proposed debarment to the investigator (45 Code of Federal Regulations sections 76.311, 76.313, 1999). The investigator-respondent has 30 days following the receipt of the notice to respond in person, in writing, or through a representative, with information and/or arguments related to the proposed debarment. If the official who is conducting the debarment determines from these materials that there is a genuine dispute with respect to facts that are material to the proposed debarment, the investigator will be permitted to appear with a representative and witnesses, to provide additional documentary evidence, and to confront the witnesses against him or her.

The official conducting the debarring proceeding must make a decision within 45 days after the receipt of any documentation and argument by the investigator, unless the official finds good cause to grant an extension. In cases where there are facts that are in dispute, the debarring official may refer the case to another official for review and resolution. If this is done, the original debarring official must accept the conclusions of the second official with respect to the disputed facts unless the debarring official determines that the conclusions are either clearly erroneous or are arbitrary and capricious (45 Code of Federal Regulations section 76.314(b), 1999). In order for an investigator to be debarred, the case against him or her must be established by the agency bringing the action by a preponderance of the evidence. (See Appendix 2 for an explanation of the legal burden of proof.) If the debarment action is based on a conviction or a civil judgment against the investigator, this requirement will be deemed fulfilled (45 Code of Federal Regulations section 76.314(c), 1999).

The debarring official must provide the investigator-respondent with prompt notice of the decision to debar or not. If the official decides in favor of debarment, the period for which an investigator may be debarred must be proportionate to the seriousness of the offense. Frequently, a debarment will be imposed for a period of three years. However, the official may decide to extend this if he or she believes that it is in the public interest to do so (45 Code of Federal Regulations section 76.320, 1999).

One alternative to debarment as a sanction is that of suspension. The procedures in cases of potential suspension are similar in many respects to those in debarment actions. A suspension may precede a debarment and is generally imposed when it is believed that immediate action is necessary to protect the public interest (45 Code of Federal Regulations sections 400, 410-413, 1999). A suspension, though, can be imposed only for a temporary period pending the completion of an investigation or the initiation of other proceedings, such as a debarment proceeding (45 Code of Federal Regulations, section 76.415, 1999). In general, the suspension must be terminated with 12 months from the date of the issuance of the notice of suspension of another action, such as a debarment, has not been initiated. No suspension, however, may be imposed for a term exceeding 18 months (45 Code of Federal Regulations, section 76.415, 1999).

Exercise

You are a member of an IRB at an academic institution. The IRB approved a protocol to allow investigators to study predictors of the use of crack cocaine among women. It has come to the attention of the IRB, through the report of a co-investigator, that many of the women who are enrolled into the study are "high" at the time that they are given information in accordance with the informed consent procedure.
1. What other information do you need at this time, if any?
2. Is there an ethical breach here? If so, what is it? Refer to theories, principles, and rules in your response.
3. What courses of action are open to you and the other members of the IRB? Which will you pursue and why?

SCIENTIFIC MISCONDUCT

Defining Scientific Misconduct

Scientific misconduct has been defined by the United States Department of Health and Human Services as

> fabrication, falsification, plagiarism, or other practices that seriously deviate from those that are commonly accepted practices within the scientific community for proposing, conducting, or reporting research. It does not include honest error or honest differences in interpretations or judgments of data. (42 Code of Federal Regulations section 50.102, 1999)

Plagiarism, in turn, has been defined as including

> both the theft or misappropriation of intellectual property and the substantial unattributed textual copying of another's work. It does not include authorship or credit disputes. Substantial unattributed textual copying of another's work means the unattributed verbatim or nearly verbatim copying of sentences and paragraphs which materially mislead the ordinary reader regarding the contributions of the author. (ORI Newsletter, 1994).

In contrast, the National Science Foundation (NSF) has defined misconduct as

> (1) fabrication, falsification, plagiarism or other serious deviation from accepted practices in the proposing, carrying out, or reporting results from activities funded by NSF; or (2) retaliation of any kind

against a person who reported or provided information about suspected or alleged misconduct and who has not acted in bad faith. (45 Code of Federal Regulations section 689.1, 1998).

Other definitions have been proposed in the United States, but have not been implemented. For instance, the Commission on Research Integrity of the United States Department of Health and Human Services had proposed in 1998 that the term "misconduct in science" be replaced with "research misconduct," which would be defined as

> significant misbehavior that improperly appropriates the intellectual property or contributions of others, that intentionally impedes the progress of research or risks corrupting the scientific record or compromising the integrity of scientific practices. Such behaviors are unethical and unacceptable in proposing, conducting, or reporting research, or in reviewing the proposals or research reports of others.

One commentator (Benson, 1991a) has noted the difficulty in categorizing behaviors as "misconduct" when they fall outside of fabrication, falsification, and plagiarism. Both NSF and HHS have interpreted their respective regulations to include the falsification of credentials, such as degrees (ORI, 1994a, b, c, d,) and the falsification of letters of recommendation (ORI, 1995). Benson, citing Binder (1989), suggested the arrangement of scientific "sins" into a "concentric circle of unacceptable practices." The most unambiguous cases of fraud include plagiarism, falsification, and fabrication; these constitute the ninth through the seventh concentric rings. Those who fail to credit others for their work by omitting citation fall into the sixth circle. The fifth through the first circles encompass, respectively, scientists who selectively report data, who use inappropriate statistics, who trim data to massage the results, who use historical controls instead of actual controls for an experiment, and who divide a scientific project into the least divisible unit to produce the greatest number of scientific papers possible. A recent focus group based study conducted with National Science Foundation-funded scientists and institutional representatives found general agreement that the fabrication of data constitutes the "worst form of data inaccuracy," followed by the promulgation of inaccurate data (Wenger, Korenman, Berk, and Berry, 1997: 374). The misappropriation of others' data and the failure to attribute credit to the work of others were also seen as violations of ethics.

Onek (1994) has questioned the propriety of equating scientific misconduct with "other serious deviations from accepted research practices," which may be detrimental to the research process, but which do not constitute scientific misconduct. A related issue is that of intent, and whether researchers who have negligently engaged in harmful conduct should be subject to the same sanctions as those who are aware of the nature and implication of their acts (Dresser, 1993a, b). It should be noted, too, that scientific misconduct, as defined above, does not encompass such actions as multiple publication of the same material or authorship disputes, which often must be resolved within or between institutions and editorial boards (see Nigg and Radulescu, 1994).

The United States' definition of scientific misconduct is not universal. For instance, Denmark defines scientific dishonesty as "intention or gross negligence leading to falsification or distortion of the scientific message or a false credit or emphasis given to a scientist" (Nylenna et al., 1998: 58, Table 1; see Riis, 1993). Finland's definition approximates that of the U.S.: "presentation to the scientific community of fabricated, falsified, or misappropriated observations or results and violation against good scientific practice." Norway classifies as scientific dishonesty "all serious deviation from accepted ethical research practice in proposing, performing and reporting research." Sweden considers scientific dishonesty to be "intention [sic] distortion of the research process by fabrication of data, text, hypothesis, or methods from another researcher's manuscript or application form or publication; or distortion of the research process in other ways (Nyenna et al., 1999: 58, Table 1). Denmark and Sweden, unlike the United States, specifically require intent as an element of misconduct.

The literature is replete with examples of scientific misconduct. Only a few are presented here to illustrate the concept and to examine the ethical implications of such conduct.

In 1974, Dr. William Summerlin of the Sloan-Kettering Institute used a black felt-tip pen to darken a transplanted skin patch in two white mice. Summerlin claimed, on the basis of his "findings," to have developed a method for ensuring that the grafts would not be rejected. If this were accurate, it would have had significant implications for transplant surgery and for research related to cancer and immune system functioning. However, other researchers were unable to duplicate Summerlin's results. A lab technician, noticing something odd about the dark patch, rubbed it off with alcohol, ultimately prompting Summerlin's admission as to what he had done (Broad and Wade, 1982; McBride, 1974). Although Summerlin's relationship with Sloan-Kettering was severed, he continued to practice medicine as a dermatologist (Gore, 1981).

Alsabti, an oncologist, was infamous for the flagrancy of his plagiarism:

> Each passing month saw another group of Alsabti articles appear in various journals around the world. His method was simplicity itself. He would retype an already published paper, remove the author's name, substitute his own, and send the manuscript off to an obscure journal for publication. His tactics deceived the editors of dozens of scientific journals around the world. (Broad and Wade, 1982: 45).

Unfortunately, Alsabti's actions provoked a sad commentary on the culture of biomedical research:

> The exploits of Alsabti could never have occurred in a community of scientists where rigorous self-policing was the rule and instant expulsion was the automatic penalty for any form of dishonesty. Even when his methods eventually came to light, fellow researchers were reluctant to make a public issue of his cheating. Alsabti would be allowed to leave quietly, and would find a job in another laboratory where the same process would start over again.

> It was only after Alsabti's methods were described in a handful of international journals that the career of this...plagiarist came to a halt. (Broad and Wade, 1982: 38).

One must question why investigators would engage in scientific misconduct. Ignorance may play a role (Farthing, 1998), particularly among new investigators. Various writers have intimated that the desire and the pressures to succeed in a research career prompt much of the behavior. Broad and Wade (1982: 59) have observed, for instance, with respect to plagiarism that:

> The rewards in science are supposed to go strictly and exclusively for originality. That is why scientists strive so desperately to establish priority for their discoveries. It is also why, to judge from the frequency and bitterness of complaints, researchers sometimes fail to make fair acknowledgement of the work of their colleagues and competitors.

Kubie (1958) noted the role of external pressures in the process of distortion: the pursuit of money, status, or fame, and the desire to satisfy practical, commercial, or humanistic purposes. Numerous factors may contribute to the ease of engaging in misconduct: the dilution of responsibility for any particular phase of a research project, the depersonalization of the research environment, and the lack of time available to principal investigators to directly supervise staff and the direction of the research project (Kuzma, 1992). Researchers conducting focus groups with scientists funded by the National Science Foundation and with institutional representatives found that

> The conflict between the goals of research institutions and the ethics of research was felt by scientists and [institutional representatives] to contribute to ethical violations. Institutions aim to enhance their strength and prestige by maximizing research productivity. Systems that prevent, monitor, and investigate scientific misconduct consume resources and often hinder productivity. (Wenger, Korenman, Berk, and Berry, 1997: 376)

It has also been suggested that scientific fraud is a form of Munchausen's syndrome, in which individuals receive attention due to their own invented diseases (Swan, 1993). Alternatively, acts of scientific misconduct may reflect an underlying disorder whereby an individual is essentially unable to distinguish between truth and falsehood, fact and fiction (Swan, 1996).

The ethical implications of such conduct are far from minimal, regardless of one's theoretical perspective. For instance, the promulgation of fraudulent findings is unlikely, from a utilitarian perspective, to maximize good. Indeed, it may do just the opposite. Individual patients may rely on the fraudulent findings in their decision-making, to their detriment. As knowledge of the fraud becomes widespread, public confidence in the integrity of science and scientists may diminish. From a deontological perspective, such conduct reflects a lack of respect for persons. If research findings are falsified, resulting in an inability to utilize the data that was collected, individuals who have participated will have contributed their effort and time for no meaningful end. The "findings" resulting from such research may be used as a basis for decision making relating to allocation of resources in both the research and the clinical care contexts. Such judgments may be faulty as a result, thereby reducing the likelihood of achieving distributive justice.

The practical implications of fraud and fabrication of data have been recognized. Referring to a study of trichlorocarbonilide (TCC) and the alleged underreporting of the rat mortality rate in connection with that study (*United States v. Keplinger*, 1986), Kuzma (1992: 379, note 74) declared:

> Governmental action is predicated upon the assumption that the data submitted to it are accurately reported. Without accurate data, the government, and therefore the public, is forced to bear the risk that accurate data would not substantiate the claim that the chemical is safe.

There are few incentives to report possible scientific misconduct. In a study commissioned by the Office of Scientific Integrity in 1993 and completed in 1995, it was found that the 68 "whistleblowers" were "highly likely to experience one or more negative consequences as a result of their whistleblowing" (Lock, 1996; ORI, 1995). In one case, for example, the complainant's fellowship at her university was not renewed (Hilts, 1991). However, respondents who are alleged to have committed scientific misconduct may also suffer severe negative consequences, despite findings that they have not done so. A 1996 study published by ORI found that 60 percent of the 54 respondents involved in closed cases where no misconduct had been found had suffered adverse consequences, including loss of employment, loss of promotion, loss of salary increases, threatened lawsuits, professional ostracism, a reduction in research or support staff, and delays in processing grant applications (ORI, 1996).

Numerous proposals have been made to reduce or eliminate scientific misconduct. These include utilization of the media as a "watchdog," the incorporation of courses on scientific ethics into university and institutional curricula, the promulgation and enforcement of internal institutional rules, the establishment of better internal quality controls in laboratories, and more open sharing of data and research findings (Riss, 1994).

Institutional Responses

Primary responsibility for the conduct of an inquiry and an investigation of an allegation of scientific misconduct lies with the institution in which the research is being conducted. All individuals involved in research funded by the Public Health Service (PHS) are subject to inquiry and investigation on the basis of an allegation of scientific misconduct. This includes, for instance, students, residents, postdoctoral fellows, staff, faculty, and professional staff, as well as foreign and national institutions, regardless of where they are physically located (Ruling on Respondent's Motion to Dismiss, 1995).

Federal regulations require that the institution initiate an inquiry immediately following the receipt of an allegation. The inquiry is a preliminary investigation conducted to determine whether the allegation has sufficient substance to warrant a full investigation; it is not a procedure to reach a final conclusion about whether misconduct has occurred and who is responsible. The inquiry should be completed within 60 days. In some instances, the complainant will communicate directly with ORI regarding an allegation of misconduct. In such cases, ORI will request that the institution commence an inquiry (United States Department of Health and Human Services, 1993).

Individuals who are to be the subject of an inquiry must be notified in writing of the inquiry. They should be informed of the research project in question, the specific allegations, the definition of scientific misconduct, the PHS funding involved, and the names of the individuals on the inquiry committee and any experts. The individual should also be informed of his or her right to challenge the appointment of a committee member or expert on the basis of bias or conflict of interest, the right to be assisted by counsel and to present evidence to the committee, and the right to comment on the inquiry report. The notice should also contain a reminder of the respondent's obligations, including the obligation to maintain the confidentiality of the proceedings (Office of Research Integrity, *Model Procedures*).

During the inquiry, each respondent, complainant, and witness should have an opportunity to be interviewed. Prior to the interview, the individual to be interviewed should be provided with a summary of the issues to be discussed at the interview. Interviews should be transcribed or recorded and should be kept confidential. It has been suggested that individuals be interviewed in the following order: complainant, key witnesses, and respondent (Office of Research Integrity, *Model Procedures*). If the respondent admits that he or she committed scientific misconduct, he or she should be asked to sign a written statement. This generally provides a sufficient basis to initiate the investigation (Office of Research Integrity, *Model Procedures*).

If an investigation is to be commenced, it must be initiated within 30 days following the completion of the inquiry. The institution must advise the director of ORI of its decision to initiate the investigation on or before the date on which it commences the investigation (42 Code of Federal Regulations section 50.104(a)(1), 1999). The notification must include the name of the individual or individuals against whom the allegation has been made, the nature of the conduct as it relates to the definition of scientific misconduct, and the PHS grants involved (42 Code of Federal Regulations section 50.104(a)(1), 1999). If the institution terminates the investigation prior to completion, it must notify ORI of the planned termination and explain the reasons for this decision (42 Code of Federal Regulations section 50.104(a)(3), 1999).

An investigation is conducted in order to examine the evidence in greater depth and to determine whether scientific misconduct has occurred, to what extent, and by whom (Office of Research Integrity, *Model Procedures*). The respondent must be notified of a decision to commence an investigation. He or she should receive a copy of the inquiry report and notification of the specific allegations, the sources of PHS funding, the definition of scientific misconduct, the procedures to be followed for the investigation, and an explanation of the right to appeal to the

Departmental Appeals Board (see below) if there is a finding by ORI of scientific misconduct.

The Office of Research Integrity (ORI), described below, must be informed by the institution at any stage of the inquiry or investigation if any of the following circumstances are present:
1. there is an immediate health hazard involved,
2. there is an immediate need to protect federal funds or equipment,
3. there is an immediate need to protect the interests of the person or persons who made the allegations of scientific misconduct or the individual or individuals who are the subject of the complaint,
4. it is likely that the incident will be reported publicly,
5. the allegation involves a sensitive public health issue, or
6. there is a reasonable indication of a criminal violation (42 Code of Federal regulations section 50.104(b), 1999).

The institution must submit its investigative report to ORI within 120 days of initiating the investigation. If the institution will require more than the 120 permitted for the completion of the investigation, it must request an extension from ORI in writing. That request must detail the reasons for the delay, the progress of the investigation, the steps remaining to be taken prior to completion, and the projected date of completion (42 Code of Federal Regulations section 50.104(a)(5), 1999).

ORI has characterized a "good investigation" by an institution as one in which the allegations are stated clearly, interviews have been conducted with all persons who may have relevant information, the sequestration of data in question was effectuated on a timely basis, all relevant documentation and research data have been thoroughly reviewed, findings are supported by documentation, the team conducting the investigation was knowledgeable and objective, the final written report was well-organized and clearly written, comments from both the complainant and the respondent are included, and confidentiality was maintained (United States Department of Health and Human Services, 1999).

The burden of proof for making a finding of scientific misconduct is on the institution. The institution must establish the scientific misconduct by a preponderance of the evidence, *i.e.*, it is more likely than not that the individual committed scientific misconduct (Office of Research Integrity, *Model Procedures*).

Institutional action upon a finding of scientific misconduct may include the denial or revocation of tenure, the withdrawal of principal investigator status, the issuance of a letter of reprimand, the review of the respondent's applications, and/or the requirement that the investigator withdraw the manuscript(s) and correct the literature. Courts have specifically found that an individual does not have a constitutionally protected right to continue to serve as the principal investigator of a PHS-funded grant because institutions are the grantees of the awards (*Hiserodt v. Shalala*, 1994; *Needleman v. Healy et al.*, 1996).

Administrative Responses to Scientific Misconduct

The Office of Research Integrity

Prior to the mid-1980s, allegations of scientific misconduct were handled informally by institutions and the relevant federal agencies. However, as the result of the public disclosure of numerous instance of misconduct, Congress enacted in 1985 the Health Research Extension Act. Pursuant to this law, institutions receiving federal funding for their research were required to establish "an administrative review process to review reports of scientific fraud" and to report to the "Secretary any investigation of alleged scientific fraud which appears substantial." The National Institutes of Health published guidelines in July 1986 and final regulations appeared in the Federal Register on August 8, 1989 (United States Department of Health and Human Services, 1999). In March 1989, the Public Health Service created the Office of Scientific Integrity in the Office of the Director of NIH and the Office of Scientific Integrity Review in the Office of the Assistant Secretary for Health of the HHS. These offices were consolidated into the Office of Research Integrity in June 1992.

Institutions conducting the research have the primary responsibility for investigating allegations of scientific misconduct. Consequently, ORI's responsibility generally consists of reviewing the institution's investigative report. ORI will review conduct that relates to research funded by the Public Health Service or an application for PHS funding and that falls within the definition of scientific misconduct noted above. ORI will review the investigation to verify that it was fair and thorough and that the evidence supports the findings. Based on its review, ORI may accept or reject the findings of the institution, request further investigation, or begin its own investigation (United States Department of Health and Human Services, 1999).

The ORI will notify the researcher alleged to have committed misconduct (the respondent) of its proposed findings. If misconduct has been found, the respondent has 30 days to request a hearing before the Department Appeals Board (see below) on the findings and proposed administrative action. Action recommended by ORI in response to a finding of misconduct may include any of the following: debarment from federal funding, the imposition of a prohibition against serving on PHS advisory committees, a requirement that the institution certify the accuracy of respondent's applications, and a requirement that the respondent's research be supervised (United States Department of Health and Human Services, 1999).

Since its inception in 1992, ORI has addressed more than 1,500 allegations of scientific misconduct. Approximately 20 percent have required a formal inquiry, utilizing the procedures described below. Of the 150 cases investigated between 1993 and 1997, 76 resulted in findings of scientific misconduct and 74 resulted in findings of no scientific misconduct (Office of Research Integrity, 1999; see Office of Research Integrity, *Summaries*). Of the cases in which misconduct was found, falsification was the most frequent type of misconduct, followed by fabrication and then by plagiarism. The most frequent action taken by the investigators' institution was a reprimand. A total of 170 administrative actions were imposed on the 76 respondents found to have committed misconduct. These include debarment from the receipt of federal funds for a period of time ranging from 18 months to 8 years (71%), a prohibition from serving on a PHS advisory panel (91%), a requirement that research be supervised (26%), a requirement that data be certified (13%), a requirement that sources be certified (9%), and a correction or retraction of articles (13%) (Office of Research Integrity, 1999).

The Departmental Appeals Board

Findings of the Office of Research Integrity can be appealed to the Departmental Appeals Board (DAB). The Chair of the DAB will appoint a Research Integrity Adjudication Panel, composed of administrative law judges, DAB members, and scientists (United States Department of Health and Human Services, 1999). The ORI is represented by the Research Integrity Branch of the Office of the General Counsel in hearings before the DAB. In hearings before the DAB, the ORI must establish that the respondent committed scientific misconduct by a preponderance of the evidence (*John C. Hiserodt*, 1995). Respondents can be represented by an attorney at these hearings and have the right to question any evidence and witnesses presented by ORI. They may also present witnesses and evidence to rebut the findings and the proposed administrative action (United States Department of Health and Human Services, 1999). The decision of the DAB will be made after hearing all of the evidence and witnesses presented by the respondent and by ORI. The decision of the DAB is the final PHS decision. The DAB may review the sanctions imposed by ORI, such as debarment from federal funding, the imposition of a prohibition against serving on PHS advisory committees, a requirement that the institution certify the accuracy of respondent's applications, and a requirement that the respondent's research be supervised (United States Department of Health and Human Services, 1999).

The Angelides Decision (DAB, 1999) provides an example of ORI and DAB Procedures and the imposition of debarment. Angelides was a scientist employed at the Baylor College of Medicine in Texas. His department chairman noticed inconsistencies in grant applications that Angelides submitted to the NIH. A preliminary inquiry was held to examine possible scientific misconduct, as is required by the regulations. The first inquiry did not lead to a full investigation. However, subsequent questions were raised, leading to a second inquiry and investigation. During this investigation, Angelides acknowledged that elements of his grant applications were false. He appealed the investigation committee's conclusion that he had engaged in scientific misconduct to an appellate committee, comprised of leading scientists. That committee affirmed the finding. Angelides was dismissed from Baylor.

Angelides then filed a lawsuit against Baylor, the members of the investigation committee, and several witnesses, claiming that he had been defamed and wrongfully terminated from his position at Baylor. Concurrently, ORI conducted an oversight review of the investigation. Based on this review, ORI recommended that Angelides be subject to debarment for a period of five years and that he be subject to various other administrative sanctions. Angelides appealed the decision of ORI to DAB.

Angelides argued that the falsifications stemmed from falsifications of data provided to him by his graduate students and a postdoctoral fellow. DAB flatly rejected this argument. DAB also found Angelides guilty of scientific misconduct. In so doing, DAB provided insight into the standards that should govern scientific research:

> Dr. Angelides's own expert witness testified that, if a manuscript is prepared after a student has left a laboratory, the standard procedure would be to contact the experimentalist to review and interpret the primary data and participate in the preparation of the manuscript. They agreed that the standards in the scientific community, then as now, required a good faith effort to ensure accurate reporting of others' data. In fact, Dr. Angelides himself agreed that he had an obligation to consult his students and resolve any questions about molecular weight or tissue source. Other scientists agreed that the interpretation of data should be verified with the person who conducted the experiment if the data are not labeled clearly enough to preclude error....Dr. Angelides had the primary responsibility for the accuracy of the data, since in each case he provided them for publication without seeking the review and input of the actual experimentalists. (DAB, 1999: 113).

The DAB also rejected Angelides' argument that false statements on the grant applications should not be construed as scientific misconduct because they were not critical to the overall conclusions of the papers or to the decision to fund the grant applications. The DAB stated (1999: 65):

> Such a proposition would permit a scientist, with impunity, to knowingly make false claims that overstate the capabilities or achievements of a laboratory, as compared to others that may also be seeking funding in a very competitive funding environment, so long as the misrepresentations in a particular grant were not about the grant's central project or so long as the scientist could suggest alternative approaches to making the falsified data "optional." The integrity of the funding process, which depends on accurate and honest information, could be undermined....Hence, any statement included in a grant application that portrays to reviewers the capacities or accomplishments of the researcher or laboratory as further advanced than they are in reality, or presents a more favorable picture of the likelihood of success than the true facts would suggest, can therefore be considered as material to the funding decision, whether or not it was "necessary" to the presentation of the research proposal. At the same time, it is evident as a general proposition that the more favorable and the more significant the false statement is, all other things being equal, the greater the likelihood that the misrepresentation is intentional.

The following day, the civil lawsuit against Baylor and the other defendants was settled and Angelides agreed not to appeal DAB's decision.

The current federal process for responding to allegations of scientific misconduct and, in particular, the DAB, have been subject to a variety of criticisms. First, members of the DAB, often lack a scientific background and, consequently, have difficulty understanding both the underlying scientific principles and the ethos of the scientific community (Parrish, 1997). Second, many IRB members do not attend the majority of the hearings to which they have been assigned, so that they are unable to ask questions of the expert witnesses and are unable to evaluate the credibility of the witnesses whose testimony they have not seen or heard. And, finally, some critics have charged that DAB lacks the necessary skills to conduct what is essentially an adversarial de novo proceeding, and not simply a review of the evidence that was resented to ORI (Parrish, 1997).

Exercise

You are the principal investigator of an interview-based study that seeks to examine individuals' perceptions of what constitutes elder abuse and neglect. All interviews have been tape recorded. Participants are paid a small stipend to thank them for their time, since the interviews are quite lengthy. It has come to your attention through "the grapevine" that, rather than utilizing the recruitment scheme that had been designed for the study and approved by the IRB, the interviewers have been interviewing their friends.

1. What additional information, if any, do you need at this time?
2. What courses of action are open to you as the principal investigator? Which would you select and why?
3. What, if any, harm has occurred as the result of the interviewers' use of their friends for these interviews?

MISCONDUCT IN REGULATED RESEARCH

The Food and Drug Administration (FDA) is responsible for the promulgation of regulations for the protection of human subjects participating in clinical investigations regulated by the FDA and clinical applications for research or marketing permits for products regulated by the FDA. These include food and color additives, drugs for human use, and electronic products (21 Code of Federal Regulations section 50.1, 1998). A "clinical investigation" refers to any experiment that involves a test article and at least one human participant, and is either subject to certain FDA submission requirements, or the results of which will be submitted to the FDA in conjunction with an application for a research or marketing permit (21 Code of Federal Regulations section 50.3, 1998). These regulations may be applicable regardless of whether or not the clinical investigation is funded by the FDA.

FDA regulations require that studies involving investigational new drugs, medical devices, and biologics that are conducted with human subjects be reviewed prior to their initiation by an institutional review board. The regulations specify that

FDA may inspect IRBs and review and copy IRB records (21 Code of Federal Regulations section 56.115(b), 1998). This function derives from its Bioresearch Monitoring Program, which the FDA established in 1977 to "ensure the quality and integrity of data submitted to FDA for regulatory decisions, as well as to protect human subjects of research" (FDA, 1998). These IRB reviews are conducted in order to determine if an IRB is operating in accordance with FDA regulations relating to IRBs and with the IRB's own written procedures.

The Bioresearch Monitoring Program encompasses three types of inspections: investigator-oriented inspections, study-oriented inspections, and bioequivalence study inspections. Only the first two types are discussed here.

Study-oriented inspections focus on studies that are important to product evaluation. Examples include new drug applications and product license applications. An investigator-oriented inspection may be initiated for any of the following reasons:

1. The investigator conducted an extraordinarily important study that has particular significance with respect to medical practice or product approval.
2. Representatives of the research sponsor, such as a pharmaceutical company, have reported difficulties in getting case reports from the investigator.
3. Representatives of the research sponsor have reported some concerns with regard to the investigator's work.
4. A participant in a study complained about protocol or human subjects violations.
5. The investigator has participated in a large number of studies or has done work outside his or her specialty area.
6. Safety or effectiveness findings are inconsistent with those of other investigators who have studied the same test article.
7. The investigator has claimed too many subjects with a specified disease relative to the location of the investigation.
8. Laboratory results are outside of the range of expected biological variation.

The procedures for study-oriented inspections and investigator-oriented inspections are similar. A representative of the FDA District Office will contact the researcher under investigation to arrange a meeting. The investigation at that meeting will initially focus on the relevant factual circumstances, such as the allocation of responsibility and degree of delegation of authority, various aspects of the study, and the procedures used to collect and record data. The FDA representative will then compare the data submitted to the FDA and/or the research sponsor with any available records that might support the data (FDA, 1998).

The FDA representative will conduct an exit interview with the clinical investigator at the end of the inspection. The representative will discuss the findings resulting from the inspection, clarify any misunderstandings and, in some cases, will issue a written Form FDA-483, which is entitled Inspectional Observations. The representative will then prepare a written report and will submit it to headquarters for evaluation.

After the report is evaluated, one of three types of letters will be issued to the investigator:

1. The letter will state that there were no significant deviations noted. This type of letter does not require that the clinical investigator respond.
2. An informational letter will identify any deviations from regulations and from good clinical practice. In some cases, a response will be required from the clinical investigator. If this is expected, the letter will detail what must be done and provide the name of a contact person should the investigator have any questions.
3. A warning letter will be issued, which identifies serious deviations from the relevant regulations. This type of letter requires an immediate response from the clinical investigator. The FDA will inform the sponsor of the study and the IRB responsible for reviewing the study of the deficiencies that were noted. If the FDA observed deficiencies in the monitoring of the study by its sponsor, the FDA will so notify the sponsor. The FDA may also impose administrative and/or regulatory sanctions in such cases.

One such sanction is that of disqualifying an investigator from receiving investigational drugs, biologics and devices. This sanction can be imposed only if it is found that the investigator has repeatedly or deliberately violated the regulations of the FDA, or if the investigator gave false information to a sponsor in a required report. In such cases, the FDA will send the investigator a written notice that advises of the noncompliance or the false submission. The investigator will be advised of a specified time period during which he or she may respond to the notice, either in writing or at a conference. Although the conference is supposed to be informal, a transcript will be made and the investigator may bring a legal representative with him or her (FDA, 1998).

If the FDA finds that the investigator's response to the written notice was both timely and satisfactory, it will advise the investigator of that and will terminate the proceeding. However, if the FDA feels that the explanation proffered by the investigator is unsatisfactory or if the investigator does not respond within the designated time interval, the FDA will offer a regulatory "Part 16" hearing to the investigator. This is designated as an informal hearing and is conducted to determine whether or not the investigator should continue to be eligible to receive investigational test articles.

The Part 16 hearing is initiated by the issuance by the FDA of a written Notice of Opportunity for Hearing. The Notice details the allegations against the investigator and sets forth the other information that is the subject of the hearing. The hearing will not be held if the investigator does not respond to the Notice within the time period designated for response in the Notice. If the investigator does respond and requests a hearing, the FDA Commissioner will designate a presiding officer from the Office of Health Affairs (OHA). The hearing will be held at the FDA headquarters.

Prior to the hearing, the investigator and the FDA Center which has sent the Notice may exchange published articles or written information that will be utilized at the time of the hearing. Copies of documents will be provided by each party to the other if it is unlikely that the other party would have a copy of such documents and the documents will be relied upon at the time of the hearing. The Center and/or the investigator may file a motion for a summary decision at this point.

The rules of evidence that apply in court trials do not apply to these hearings. The hearing is conducted by the presiding officer. The staff of the FDA Center first presents a statement which indicates the subject of the action, describes the supporting information, and explains why the investigator should be disqualified. The investigator may be represented by an attorney and may present information that is relevant.

The OHA officer will prepare a written report at the conclusion of the hearing. The administrative record of the hearing includes all of the written material that was presented at the hearing, as well as a transcript of the proceedings. The parties will be given a chance to review and comment on the written report. The Commissioner will then review the written report, the parties' comments, and the administrative record to decide if the investigator should be disqualified. The Commissioner will then issue a written decision which includes the underlying reasons for that decision.

In cases in which the investigator is to be disqualified, the Commissioner must notify the sponsor of the investigation, notify the sponsors of studies conducted under each investigational new drug (IND), investigational device exemption (IDE), or approved application that contains data from the investigator that the FDA will not accept the investigator's work without validating information indicating that it was not affected by the misconduct, determine whether those data can support the IND or IDE studies after the investigator's data have been disregarded, and determine whether the product should continue to receive approval (FDA, 1998). The Notice of Opportunity for a Hearing is available to the public through the Freedom of Information Act. (See below for a discussion of the Act.)

In some cases, the investigator may enter into a consent agreement in addition to utilizing the opportunity for an informal conference. In such cases, the disqualification process will not continue. The consent agreement will usually provide that either the investigator agrees not to conduct studies with FDA-regulated test articles or, if the investigator continues to conduct such studies, the studies will be subject to various restrictions, such as oversight by a specified individual. In some circumstances, the investigator may qualify for reinstatement.

Although a case may be resolved through the issuance of a final order in a Part 16 proceeding or through the entry into a consent agreement, this terminates only the administrative action. However, FDA may still refer the case for criminal prosecution, especially in situations in which the clinical investigator has knowingly or willingly provided false information to the research sponsor. Criminal actions are discussed in greater detail below.

Four types of misconduct have been noted from FDA audits: (1) the deliberate fabrication of results, known as dry labbing; (2) the violation of regulations governing research, such as a failure to obtain informed consent; (3) the modification of data to enhance its publishability, often referred to as fudging; and (4) the non-deliberate violation of research norms and regulations, often due to a lack of understanding of basic research principles (Horowitz, 1996).

Exercise

The IRB of which you are a member has approved a protocol for the conduct of a phase I clinical trial of a new drug which, if effective, will decrease hallucinations in individuals suffering from schizophrenia. It has come to your attention that many of the individuals who are enrolled in the study are acutely psychotic.
1. Is this a situation that requires additional investigation? Why or why not?
2. What additional information do you need, if any?
3. What courses of action are open to the IRB? Which would you pursue and why?

LEGAL RESPONSES TO MISUSE OF HUMAN PARTICIPANTS, SCIENTIFIC MISCONDUCT, AND MISCONDUCT IN REGULATED RESEARCH

This section addresses potential legal responses to the misuse of human participants, to scientific misconduct, and to misconduct in regulated research. As such, it focuses on the legal obligations that are imposed through statutes, regulations, and common law. In reviewing actual cases, it is easy to focus on the resulting liability for the researcher and to dismissively assert that anyone can be sued for anything. It is critical to remember, however, that the bases for these lawsuits stem from a concern for the party who may have been injured and represent attempts to balance the interests of both the party alleging harm and the party alleged to have caused that harm. Underlying each such legal claim is a possible violation of ethical obligations that the researcher has towards the participant, regardless of the ethical framework from which those obligations derive. For instance, each such framework would eschew the imposition of harm on experimental subjects.

A review of legal procedures and concepts, such as differences between civil and criminal actions and techniques related to the acquisition of evidence, may be found in Appendix 2. The reader may wish to consult that summary prior to reading this section to gain basic familiarity with judicial proceedings.

Civil Proceedings

This discussion focuses on injuries known as torts. There are three general divisions of torts: personal injury, property damage, and invasion of interests, such as privacy. Those that are most relevant in the context of scientific research are injuries relating to the person and injuries relating to specific interests. Within each of these categories, the conduct that caused the harm may have been intentional or it may have been negligent. In certain cases, the law imposes what is called "strict liability," regardless of the existence of intent or negligence. Only those torts within each of these categories that are most relevant to the context of scientific research are addressed here. Unlike a criminal prosecution, in which the defendant will be found guilty or not guilty and, if guilty, may be sentenced or fined criminally, a finding for the party bringing the civil lawsuit (plaintiff) against the defendant can potentially result in an award of money (damages) or in the medical monitoring of a condition, such as occurred with the lawsuits against the University of Chicago in connection with the DES experiment.

Intentional Torts

Battery occurs when a person touches another with the intent to inflict a harmful or offensive touching. Whether the person doing the touching had the intent depends on whether he or she wanted the harmful or offensive touching or whether he or she believed that such a touching was substantially certain to result from his or her act (*Frey v. Kouf*, 1992). Intent is not the same thing as motive; the motive is irrelevant. A touching is considered harmful if it results in pain, disfigurement, or impairment of a bodily organ or function. If the person who is harmed, though, gave legally effective consent, meaning that he or she understood the conduct to be done, voluntarily agreed to the touching, and had the capacity to consent, there is no intentional battery.

Assume, for instance, that an individual is enrolled in a clinical trial that is designed to evaluate the efficacy of a particular heart device. The individual suffers harm as the result of having the device implanted. If the individual had given his informed consent to participate in that trial, there would be no basis in fact for a claim that the investigator had committed an intentional battery. Assume, though, that the study "participant"-patient did not know that he was participating in this experiment, but had signed an informed consent form for a completely different study. Because of his serious condition, the patient was actually not even eligible for enrollment into the clinical trial in question because the risks to his heart were so great. In such a case, the participant could potentially sue the investigator, claiming that he or she believed that the touching (implantation of the device) was substantially certain to result in the harm that the participant suffered, because the investigator violated his or her own protocol in enrolling the individual.

In the clinical context, physicians have been found liable for intentional battery where the patient did not consent to the medical treatment provided (*Gary v. Grunnagle*, 1966). Several courts have found similarly in the context of research. In *Mink v. University of Chicago* (1978), plaintiffs sued the University of Chicago for having used them in the DES experiment without their knowledge or consent. Even though there was no direct physical harm to the plaintiffs, they claimed that they had been injured because, as a result of the researchers' conduct, they experienced mental anxiety and emotional distress due to the increased risk of cancer in their children. *Friter v. Iolab Corporation* (1992) involved a lawsuit by a patient for injuries sustained in connection with the unauthorized placement in his eye of an intraocular lens that had not yet been approved and was still under investigation by the FDA. The appellate court found that the hospital had specifically assumed an affirmative duty to ensure that study participants provided informed consent, evidenced by their signature on a form that detailed various aspects of the study. Friter had not signed such a form. The court found that absent informed consent, the physical contact to Friter could be offensive and the hospital would be liable for battery. It would appear that under such circumstances a court could find both the investigator and the hospital liable even if the research and consent protocols had been reviewed and approved by an institutional review board.

(See Freedman and Glass, 1990, discussing such liability in Canada.) In another case involving eye surgery, the patient Kus agreed to undergo cataract surgery and a lens implant based on the physician-researcher's assurances that the procedure was safe (*Kus v. Sherman Hospital*, 1995). The physician did not disclose that the lens was under investigation for safety and effectiveness and had directed his staff to remove from the IRB-approved informed consent form the paragraph indicating that the lens was under investigation. This defective informed consent form was used for all 43 patients who underwent the experimental procedure. The court found that the defendant researcher had committed battery based on the lack of consent to the physical contact. In a clinical context, if the plaintiff claims that he or she gave informed consent, but was not adequately apprised of the risks and benefits of the proposed surgery or treatment, the claim is more likely to be considered negligent rather than intentional (*Cobbs v. Grant*, 1972).

Battery can also be the basis for criminal proceedings in specified circumstances. This is addressed below in the section entitled Criminal Proceedings.

Negligence

In all lawsuits based in negligence, the plaintiff must establish causation: there was a duty owed to the plaintiff by the defendant, that the defendant breached that duty, that harm resulted to the plaintiff as a result of that breach (cause in fact), that there was a nexus between the defendant's action and the harm (proximate, or legal cause), and that damages are claimed. (See Appendix 2 for a discussion of cause in fact and proximate cause.) In some jurisdictions, this duty of care is owed only to those who could be foreseeably injured; in other jurisdictions, if there is a duty owed to anyone, it is owed to everyone. (This difference of opinion is evident in the multiplicity of interpretations of the *Tarasoff* case, discussed below in the context of disclosure of information and a duty to warn.) The following are examples of the types of conduct that may be negligent.

The failure to provide relevant information to a patient-research participant may result in liability for the researcher, as well as harm to the participant. The case of *Moore v. Regents of the University of California* (1990) involved John Moore, who had been referred to UCLA Medical Center for the treatment of hairy cell leukemia. Moore signed a standard informed consent form for his surgery; the form did not make any mention of research being performed on his excised tissue. Later, Moore relocated to Seattle, but continued to travel to UCLA on a periodic basis for what he was told were necessary follow-up visits to test his blood. He was not told that these visits were for research purposes and he paid the travel costs for these visits himself.

In 1983, Moore was requested during one of these visits to sign a form that would provide his consent to the use of his blood for research purposes. He refused to do so. The consent form that he did not sign provided that he would "voluntarily grant to the University of California any or all rights [he or his heirs] might have in any cell-line or other potential product which might be developed from the blood and/or bone marrow obtained from [him]" (*Moore v. Regents of the University of California*, 1988: 769). In March 1994, Moore's physician and his collaborator received a patent on the cell-line that was developed solely from Moore's tissue. Although Moore sued on a number of grounds, the court ultimately found that the physician had failed to obtain Moore's informed consent:

> [A] physician who is seeking a patient's consent for a medical procedure must, in order to satisfy his fiduciary duty and to obtain the patient's informed consent, disclose personal interests unrelated to the patient's health, whether research or economic, that may affect his medical judgement. (*Moore v. Regents of the University of California*, 1990: 131)

However, the court did not find that a researcher who was not also the treating physician had any duty to inform the patient that his tissue was being used for research.

Research participants may also be harmed through an invasion of their privacy. Such conduct by the researcher could potentially result in litigation. Although wrongful invasion of privacy may take several forms, those that are relevant in the research context are the public disclosure of private facts about a participant and the use of the participant's name or picture for commercial purposes. Consider the following situations.

First, assume that the researcher is conducting a study to locate the gene, or one of the genes, responsible for the development of a particular disease. In doing familial studies, the researcher establishes family pedigrees. Assume further that this is a relatively rare disease and, as with many research participants, the families are interested in the findings of this research. The researcher publishes an article in a prestigious international professional journal explaining his findings. He includes in the article the exact pedigree of a specific family which demonstrates that the parent-child relationship between several family members that had been understood to exist was, in reality, not so and that the children had been conceived with a man other than their mother's husband. Because of the rarity of the disease and the complexity and detail of the published pedigree, the family is easily identifiable.

Consider another situation in which a patient is suffering from classical symptoms of an infectious disease. The researcher asks the patient if he can take a photograph of the patient, without disclosing the purpose of the photo. The patient assumes that it will be used as part of his medical record. In fact, the researcher includes it in materials which he uses in consulting and publication, without blocking out portions of the photograph that render the patient identifiable. In one actual case, a physician was found to have publicly disclosed private facts about a patient where he released photographs of his patient's anatomy, so that there was a highly offensive disclosure and no legitimate public interest in such disclosure (*Horne v. Patton*, 1974).

Misrepresentation may be classified as an intentional or negligent tort, or one involving strict liability, depending upon the defendants knowledge and intent. Details regarding these distinguishing elements will not be reviewed here.

In general, misrepresentation occurs where the defendant has made a material misrepresentation to the plaintiff, the plaintiff has relied on that misrepresentation, and harm has occurred as a result. For instance, suppose that a researcher wishes to recruit participants for a particular trial. The researcher has not accurately represented the risks of the procedures to either the institutional review board or to the potential participants. In such a situation, a harmed participant might sue the researcher and his or her institution for harm arising from the misrepresentation.

Criminal Proceedings

Perhaps the most famous of all criminal trials for the misuse of human subjects is that of the Nuremberg Trials, at which the Nazi physicians who performed many of the experiments discussed in chapter 1 were prosecuted for their crimes under international law. There have been various well-publicized cases in which investigators have been charged with crimes based on their scientific misconduct. For instance, The National Institutes of Health had provided Stephen Breuning, a research psychologist with the University of Pittsburgh, with more than $150,000 in grants to study the effects of treating hyperactive retarded children with Ritalin and with Dexedrine. It was alleged, however, that he never actually conducted the studies for which he received the funds. Ultimately, Breuning plead guilty (*United States v. Breuning*, 1988, cited in Kuzma, 1992: 357), and was sentenced to serve 60 days in a halfway house, provide 250 hours of community service, and 5 years of probation. Additionally, he was required to repay the university $11,352 and to remain out of psychology for 5 years (Lock, 1996). Another criminal case involved Industrial Bio-Test Laboratories, Inc. which, in the 1970s, was one of the largest laboratories engaged in contract research (Shapiro and Charrow, 1985). IBT was to conduct animal toxicity studies on products that were subject to government regulation. This included, for instance, drugs and pesticides. The FDA ultimately raised questions about the validity of IBT's findings. Several of the officers plead guilty to various charges (McTaggart, 1980; Mintz, 1979). In yet another situation in which an obstetrician-gynecologist was prosecuted for making false statements to FDA, the physician plead guilty to charges that he had falsely reported that he had administered medications to over 900 patients, that he failed to give any medications to some of the patients who he had reported had received them, that he used the funding provided by the research sponsor for his own purposes, and that he falsified the records of an institutional review committee (Shapiro and Charrow, 1985).

It is important to understand that there must be a basis upon which an investigator or institution can be charged with a crime. In some instances, international law may provide the foundation for such charges, as with the Nuremberg Trials. In other instances, there will be a federal or state statute which criminalizes specified conduct, such as making false applications on applications and reports to federal funding agencies. No matter what the basis of the prosecution, in each instance the legal entity prosecuting the case, such as the district attorney's office prosecuting violations of a state statute or a U.S. Attorney's office prosecuting the violation of a federal statute, must prove various elements of the alleged crime in order to obtain a conviction.

It is impossible within the context of this chapter to review all possible bases for the criminal prosecution of an investigator for alleged scientific misconduct or misuse of human participants. Accordingly, this section provides a general discussion of various classes of crimes that could potentially be charged in connection with the misuse of human participants, scientific misconduct, and misconduct in regulated research.

Additionally, the discussion focuses on crimes within the context of United States law. The discussion reflects the possibilities of prosecution; conviction requires that all the elements of a specified crime be proven by the prosecution and that a defendant's guilt be established. (See Appendix 2 for further discussion.) It should be remembered, however, that in some situations an investigator is potentially subject to prosecution in other jurisdictions. For instance, the Nazi doctors were prosecuted under international law. If a U.S. investigator is conducting a research study in another country and violates the criminal laws of that country in the conduct of the study, the investigator may be prosecuted under the laws of that other country. As an example, an investigator might violate the customs laws of a country in importing certain equipment or drugs.

Battery

Battery has traditionally been defined as the unlawful application of force by one person upon another (*State v. Hefner*, 1930), whether or not an injury resulted from that touching. Most modern statutes, however, require that the touching have resulted in either physical injury or that the touching was intended to be or is likely to be regarded as offensive. In general, battery will be said to have occurred if the individual accused of the battery should have been aware that his or her conduct would cause the application of force to another individual. Legally effective consent is a defense to a charge of battery. To be legally effective, the consent must have been voluntarily given, the person consenting must have had legal capacity to give consent at the time that he or she gave consent, and there cannot have been any mistake as to the nature of the defendant's conduct at the time that consent was given.

Assume, for instance, that an individual is enrolled in a clinical trial that is designed to evaluate the efficacy of a particular heart device. The individual suffers harm as the result of having the device implanted. If the individual had given his informed consent to participate in that trial, there would be no basis for a district attorney to proceed against the investigator on a criminal charge of battery. Assume, though, that the study "participant"-patient did not know that he was participating in this experiment, but had signed an informed consent form for a completely different study. Because of his serious condition, the patient was actually not even eligible for enrollment into the clinical trial in question because the risks to his heart were so great. In such a case, it is possible that the district attorney might bring criminal charges against the investigator.

Murder and Manslaughter

Assume that instead of suffering injury as a result of the implanted heart device in the above hypothetical situation, the patient actually died. Could the investigator be charged with murder or manslaughter?

Many statutes define murder as the unlawful killing of another human being with malice aforethought. Manifestations of "malice aforethought" that are relevant to this discussion are an intent to kill the victim, an intent to inflict great bodily injury, or the commission of conduct where there is an unusually high risk that the conduct will cause death or serious bodily injury. Traditionally, to constitute murder on this last basis, it has been said that the risk was so great that ignoring it demonstrates an "abandoned and malignant heart" or a "depraved mind" (*Commonwealth v. Malone*, 1946; New York Penal Law section 125.25).

Consider, for instance, the hypothetical involving the clinical trial of the heart device. The principal investigator-physician wrote the protocol and established the inclusion and exclusion criteria for the study, as well as the informed consent procedures. The study, as it was designed, was approved by an institutional review board. Then the physician-researcher violated the very protocol which he had written, recruiting and enrolling patients into the trial who, under the protocol which he had written, were at excessively high risk for participation because the risks associated with such cases outweighed the potential benefits. He failed to inform the patients that the procedure which they were to undergo was actually an experiment, and then a patient died. There is, arguably, a basis for the initiation of a criminal prosecution against the researcher for both murder and battery.

The researcher could also potentially be charged with involuntary manslaughter. Involuntary manslaughter is an unintended killing that results from criminal negligence. In order for there to be criminal negligence, there must be a high and unreasonable risk of death to the individual (*Commonwealth v. Aurick*, 1941; *Commonwealth v. Wolensky*, 1944) and, in some jurisdictions, the defendant must have been aware of the risk.

Federal Statutes

Prosecution is possible under a number of federal statutes relating to the submission of false statements to the federal government and defrauding someone through a federally controllable means, such as the mail system (18 United States Code sections 1001, 1341, 1999). For instance, the falsification of credentials in connection with a grant application may be prosecuted under a federal statute which prohibits the knowing or willful falsification of a material fact in an application submitted to the federal government (18 United States Code section 1001, 1999). Alternatively, it could be prosecuted under the False Claims Act, which prohibits making false representations to the federal government when submitting a claim for money (18 United States Code section 287, 1999). The prosecution of Breuning, described above, was initiated based on the alleged violation of a federal statute that provides for punishment in connection with the obstruction of an agency proceeding (18 United States Code section 1505, 1988). IBT had been charged with violations of the mail and wire fraud statutes (Kuzma, 1992).

Exercise

A researcher has been recruiting for a large clinical trial on a nationwide basis through advertisements on the radio. The trial is to test a new product that will halt the progressive loss of hair in middle-aged men. The informed consent forms that the individuals receive prior to enrollment make it clear that this is an experimental treatment that has not yet been approved and that is under investigation. The form also lists the risks and benefits that may be expected from participation. The investigator does not, however, indicate that the substance that he is testing was banned in a foreign country due to various adverse effects. This was also not made known to the FDA.

1. What, if any, violations of law have occurred here?
2. What recourse, if any, may be available to the participants who suffer injury?
3. Is this a situation in which the FDA could or should be involved? Why or why not?
4. What ethical violations, if any, have occurred here?

RELEASING DATA IN A LEGAL CONTEXT

This section discusses the disclosure of information about research participants that is either legally mandated or legally permitted. Consequently, the level of confidentiality that a researcher may wish to ensure to his or her research participants may be limited by various provisions in state and federal law. The most common of these are discussed below: mandated reporting laws, subpoenas from courts, requests for information under the federal Freedom of Information Act or its state counterparts, the duty to warn, and partner notification laws. This section also suggests strategies that may be used to enhance confidentiality.

Limits on Confidentiality

Mandated Reporting Laws

All states require the reporting of certain kinds of health events. For instance, all states require that health care providers report to the police gunshot wounds and/or wounds that appear to have been inflicted in a violent manner, such as a knife wound. All states require that health care providers report to public health authorities—usually either the local public health department or the state health department—diagnoses of sexually transmitted diseases, such as syphilis or gonorrhea. All states require the reporting of AIDS, while some also require that a diagnosis of HIV be reported. All states require the reporting of child abuse and neglect and many states require that specified persons report elder abuse and neglect. These reporting requirements give rise to four questions that each researcher and member of the research team must ask: (1) Is the health event one that must be reported under state law? (2) Is the person who is being affected covered under the state law? (3) Am I a mandated reporter and, if so, what procedures must be followed? (4) Am I permitted, but not required to report? If so, what should be the preferred course of action?

It is not possible within the scope of this chapter to explore these questions for each type of health event that might be required to be reported. We will address these issues, then, as they relate to elder abuse. Assume, for the purpose of this hypothetical, that you are part of a research team that is seeking to identify predictors of reduced stress among caregivers of elderly persons with Alzheimer's disease. You are confronted with a situation involving one of your participant-teams (caregiver and Alzheimer's-affected relative) in which the relative appears to be afraid of the caregiver, is left alone for long periods of time, appears quite disoriented, and is losing a worrisome amount of weight. You suspect that the caregiver is ignoring the needs of the relative, perhaps due to competing demands on her time and financial resources from her children. You must decide now what to do. Assume for the purposes of this situation that your state has a mandatory reporting law for elder abuse and neglect.

First, you must determine if, assuming that, indeed, the relative is being neglected, that neglect constitutes neglect within the meaning of your state statute. This task may appear easier than it actually is. Consider, for instance, the following.

Ohio law defines neglect as "the failure of an adult to provide for himself the goods or services necessary to avoid physical harm, mental anguish, or mental illness or the failure of a caretaker to provide such goods or services" (Ohio Revised Code Annotated section 5101.60(K), 1994). The statute appears to presume that physical harm, for instance, is either present or absent, rather than viewing it on a continuum and providing criteria by which to assess the degree of physical harm necessary to constitute neglect. Pennsylvania considers a failure by the elder or the caregiver to provide goods or services to be neglect only if those goods or services are "essential to avoid a clear and serious threat to physical or mental health" (Pennsylvania Statute Annotated, title 35, section 10225.103, 1999). Pennsylvania law provides that:

> No older adult who does not consent to the provision of protective services shall be found to be neglected solely on the grounds of environmental factors which are beyond the control of the older adult or the caretaker, such as inadequate housing, furnishings, income, clothing or medical care. (Pennsylvania Statute Annotated, title 35, section 10225.103, 1999)

Ohio's, and other states' failure to exclude uncontrollable environmental factors from consideration in assessing neglect may result in the imposition of a responsibility which cannot realistically be fulfilled. Florida defines neglect as

> the failure or omission on the part of the caregiver or disabled adult or elderly person to provide the care, supervision, and services necessary to maintain the physical and mental health of the disabled adult or elderly person including, but not limited to, food, clothing, medicine, shelter, supervision, and medical services, that a prudent person would consider essential for the well-being of a

disabled adult or elderly person. The term "neglect" also means the failure of a caregiver to make a reasonable effort to protect a disabled adult or an elderly person from abuse, neglect, or exploitation by others.

Florida is one of the few states to specify the frequency with which an act or omission must occur to constitute neglect: "Neglect" is repeated conduct or a single incident of carelessness which produces or could reasonably be expected to result in serious physical or psychological injury or a substantial risk of death (Florida Statute Annotated section 415.102(22), 1999). So, depending on what state you and the participants are located in, you may be required to assess, at least on a preliminary basis, whether the goods and services that are not being provided are necessary to avoid specified types of harm, or whether the inadequacy is due to environmental factors beyond the control of the caregiver.

Now, having decided, for the purpose of this example, that the actions by the caregiver towards the relative may constitute neglect as it is defined by your state, you must determine whether the relative is actually covered by the state law as an "elder." This is because different states define an elder differently. For instance, Nevada's statute protects those who are 60 years of age or older (Nevada Revised Statute section 200.5092, 1987), while Texas law encompasses persons 65 years of age or older (Texas Human Resources section 48-002, 1999). Connecticut's statute requires that the older person be 60 years of age or older and a resident of that state (Connecticut General Statute Annotated section 17b-450, 1998), while Pennsylvania law seeks to protect all those 60 years of age and older who are within its jurisdiction (Pennsylvania Statute Annotated, title 35, section 10225.103, 1999). Florida restricts its statutory coverage to those who are 60 years of age and older who are "suffering from the infirmities of aging as manifested by advanced age or organic brain damage, or other physical, mental, or emotional dysfunctioning to the extent that the ability of the person to provide adequately for the person's own care or protection is impaired" (Florida Statute Annotated section 415.102(12), 1999). Neither the statute nor case law provides an understanding of what constitutes "suffering from the infirmities of aging as manifested by advance age" or what constitutes adequate provision for one's own care and protection.

Assume that you have decided that the affected relative is, indeed, covered by the relevant state statute based on her age and her residence in that state. You must now decide if you are required by law to report the situation. This, too, varies by state. As an example, Nevada specifies that the following persons are obligated to report where their knowledge is acquired in the context of their professional or occupational capacities: physicians, dentists, dental hygienists, chiropractors, optometrists, podiatric physicians, medical examiners, residents, interns, professional practical nurses, physician's assistants, psychiatrists, psychologists, marriage and family therapists, alcohol or drug abuse counselors, drivers of ambulances, advanced emergency medical technicians or other persons providing medical services licensed or certified to practice [in Nevada] who examine, attend, or treat an older person who appears to have been abused, neglected, exploited or isolated; personnel of hospitals or other institutions; employees of agencies that provide nursing in the home; employees of the department of health services, of law enforcement, of facilities providing care for older persons, and of funeral homes or mortuaries; social workers, and coroners (Nevada Revised Statute section 200.5093(f), 1987). Now, if you are a health care provider who is acting as a researcher, but you are not providing patient care, the question as to whether you are a mandated reporter under Nevada law is unresolved. Alaska similarly limits the scope of its mandated reporting to those acquiring the knowledge "in performance of their professional duties" (Alaska Statute section 47.24.010, 1998), resulting in the same lack of clarity.

Ohio mandates reporting by the following: attorneys, physicians, osteopaths, podiatrists, chiropractors, dentists, psychologists, nurses, senior service providers, peace officers, coroners, clergymen, and employees of ambulatory care facilities, community alternative nursing homes, and community mental health facilities. The statute also requires reporting by those "engaged in" social work or counseling. The reporter must have "reasonable cause" (Ohio Revised Code Annotated section 5101.61(A), 1994). It is unclear from the statute whether individuals who are unlicensed in either social work or counseling but who provide services similar to a social worker, such as volunteers at a community service organization, are mandated reporters. Additionally, the statute fails to specify whether the reporting obligation arises only in the context of one's professional duties or also attaches to personal interactions. Accordingly, if you are not a social worker, but you are conducting interviews as a part of this study that are similar to what a social worker might be doing during an intake, and you are providing referrals to research participants who would like to be referred for social or supportive services, the question is whether you are "engaged in" social work, even though that is not your profession and not your specific task.

Subpoenas

A subpoena is an order from a court or administrative body to compel the appearance of a witness or the production of specified documents or records. A subpoena must be distinguished from a request to produce a document or record, which is issued by a party to litigation. This section is concerned with the court-ordered production of documents or records.

A subpoena can be issued by a court or by an administrative body with subpoena power, at either the state or the federal level. The information sought to be obtained through the subpoena may be deemed important to an investigatory proceeding, or to the conduct of a criminal or civil proceeding. The issuance of subpoenas against researchers has become increasingly common (Brennan, 1990). The following examples are illustrative.

The case of *In re Grand Jury Subpoena Dated Jan. 4, 1984* (1984) involved a waiter who was a doctoral candidate at a university. He was writing a dissertation relating to the sociology of the American restaurant. During the course of his investigation, he gathered information from a variety of sources and guaranteed confidentiality to all of them. He routinely recorded his observations and conversations in a book of field notes, which would be used to prepare his dissertation. Following a fire at the restaurant, the federal grand jury ordered the waiter to produce his notes. The waiter moved to quash (nullify) the subpoena, and claimed a scholar's privilege to maintain the confidentiality of the information. The court ruled against him, and found that the

> application of a scholar's privilege, if it exists, requires a threshold showing consisting of a detailed description of the nature and seriousness of the scholarly study in question, of the methodology employed, of the need for assurances of confidentiality to various sources to conduct the study and of the fact that the disclosure requested by the subpoena will seriously impinge on that confidentiality. (*In re Grand Jury Subpoena Dated Jan. 4, 1984*, 1984)

In a much later case relating to health research, rather than sociological research, a tobacco company requested the discovery, through the issuance of a subpoena, of data, tapes, questionnaires, medical records, death certificates, and other information that was part of ongoing medical research at a hospital. The tobacco company sought this material in connection with a claim that had been filed against it by the widow of an individual who had died from cancer. The 18,170 individuals who had participated in the research had requested and received assurances of confidentiality in exchange for their participation in the research. The compilation of the requested data would have required an expenditure of over 1,000 hours of time by the researchers. The hospital and the other parties seeking to quash the subpoena argued that forced compliance with the subpoena would also impinge on their academic freedom. The court granted the motion to quash the subpoena, after balancing the hardship of complying with the order against the need for the information, including the fact that neither the researchers nor the hospital were parties to the underlying litigation (*Application of J.R. Reynolds Tobacco Company*, 1987).

R.J. Reynolds, together with other tobacco companies, later applied for a subpoena from a federal district court in New York. Data were again sought in connection with numerous product liability lawsuits that had been filed against the tobacco companies. Again, neither the hospital nor the researchers were parties to the underlying investigations. The federal court ruled against the hospital and the researchers and ordered them to produce the information requested. The researchers and the hospital appealed to the federal circuit court, which affirmed the lower court. The companies' request was fashioned somewhat more narrowly, and sought only computer tapes and information necessary to interpret those tapes, rather than all of the raw data. However, the confidentiality of the research participants was not completely protected. The lower court order had allowed the researchers and the hospital to purge the following information from the data: names, street addresses, towns or villages, social security numbers, employers, and union registration numbers. However, the order would not allow the removal of counties of residence, union local data, and dates of birth and death, although this information could be used by the tobacco companies to identify specific individuals (*Application of American Tobacco Company*, 1989; Holder, 1989).

Research records have also been subpoenaed in cases involving toxic shock syndrome (*Farnsworth v. Proctor & Gamble Company*, 1985), DES (*Deitchman v. E.R. Squibb and Sons, Inc.*, 1984); *Andrews v. Eli Lilly & Company*,

1983), and in other cases involving tobacco (Barinaga, 1992; Holder, 1993). Once material is obtained via a subpoena, it is generally open to public inspection.

Requests Under the Freedom of Information Act and Similar State Statutes

The Freedom of Information Act (FOIA) is a federal statute that provides for the public inspection and copying of specifically enumerated types of information from federal agencies. This includes the opinions of federal agencies, administrative staff manuals, and policies and interpretations that have not been published in the *Federal Register*, which is a government publication in which all interim and final regulations are published. More importantly in the context of research, FOIA provides for access to "records" held by a federal agency and a general index of such records, on an individual's written request. This may include federally sponsored or funded research and the corresponding data on individuals. The agency from which the records are requested may charge reasonable, standard fees for document search, or supplication, depending on the nature of the requesting entity (5 United States Code Annotated section 552(a)(4) 1996).

The statute specifies that the agency from whom the information is requested must determine within 20 days of receiving a request whether it will comply, and must advise the requester of the information of its intent to comply or to refuse to comply. An individual whose request for information is denied has the right to file an administrative appeal from this decision. The determination of an appeal from the initial decision must be made within 20 days, excluding weekends and holidays, following the receipt of the appeal (5 United States Code Annotated section 552(a)(6)(A), 1996).

The statute provides that "[u]pon any determination by an agency to comply with a request for records, the records shall be made promptly available to such person making such request" (Freedom of Information Act, 5 United States Code section 552(a)(6)(C), 1996). However, Congress has not allocated sufficient resources to most agencies to allow them to do this. Consequently, an individual may have to wait a protracted amount of time to receive the response to his or her request (Sinrod, 1994). If the request for the information is denied, and the individual is again unsuccessful with his or her administrative appeal, he or she has the right to bring an action in federal court to attempt to compel disclosure of the information sought (Freedom of Information Act, 5 United States Code Annotated section 552(a)(4)(B), 1996).

FOIA specifically protects nine classes of information from disclosure through this procedure:
1. information that is classified or to be kept secret by Executive Order,
2. information that relates to the internal personnel rules and practices of the agency,
3. information specifically exempted by statute,
4. trade secret or other commercial or financial information that is obtained from a person and that is privileged or confidential,
5. inter- or intra-agency memoranda or letters that would not normally be available to individuals outside of a litigation context,

6. personnel and medical files, where disclosure would be "a clearly unwarranted invasion of personal privacy,"
7. specifically enumerated types of information relating to law enforcement,
8. information relating to agencies responsible for the regulations or supervision of financial institutions, and
9. geological or geophysical information and data (Freedom of Information Act, 5 United States Code Annotated section 552(b), 1996, 1999).

The definition of "agency record" as delineated in the statute is somewhat circular. "Agency" is defined as including

> any executive department, Government corporation, Government controlled corporation, or other establishment in the Executive branch of the Government (including the Executive Office of the President), or any independent regulatory agency (5 United States Code Annotated section 552(f)(1), 1999).

The provision relating to "record" states that

> "record" and any other term used in this section in reference to information includes any information that would be an agency record subject to the requirements of this section when maintained by an agency in any format, including an electronic format. (5 United States Code Annotated section 552(f)(2), 1999).

Prior to the incorporation of this definition into the statute in 1996, the courts had ruled that documents are not considered "agency records" based on a transfer from a non-FOIA agency to FOIA agency (*Kissinger v. Reporters Committee for Freedom of the Press*, 1980) or because their creation was financially supported by a FOIA agency (*Forsham v. Harris*, 1980). The Supreme Court had enunciated two prerequisites essential for the classification of requested materials as "agency records," which may still prove relevant: the agency must either "create or obtain" a record, and the agency must be in control of the requested materials at the time that the FOIA request is made (*Department of Justice v. Tax Analysts*, 1989).

Despite this ambiguity, it is clear that some information about a specific study may be available from its very inception. The submission of a funding proposal creates an "agency record" that can be accessed through reliance on a FOIA request (*Washington Research Project v. Department of Health, Education and Welfare*, 1974). This allows a requester to review the submission, whether or not it was ultimately funded. Raw data held by private grantees of federal funding are not, however, "agency records" for the purpose of disclosure under FOIA (*Forsham v. Harris*, 1980).

The use of FOIA to gain access to data became a focal point in "The Case of the Florida Dentist." Richard Driskill, a 31-year old citrus worker, claimed that he had contracted HIV from his dentist David Acer. CIGNA Dental Health of Florida, the dental program that provided Acer's services, and two experts hired by

CIGNA, obtained data from the Centers for Disease Control and Prevention (CDC) through FOIA. Using those data, they prepared their own molecular analysis and a critique of CDC's procedures, used as the basis for CDC's conclusion that Driskill may have contracted HIV while receiving dental care from Acer. The experts' receipt of data through FOIA aroused some controversy:

> Eaton, Driskill's lawyer, complains that while it may have been legal for the researchers to use the FOIA to obtain the CDC's data, they behaved unethically. "If you take someone else's work, and you don't ask permission to use it, that's wrong." However, Barbara Mishkin, an attorney for the Washington firm of Hogan and Hartson and an expert on scientific ethics—who is not involved with this case—says that data gathered by government is fair game, "especially when it forms the basis for public policy." (Palca, 1992)

Many states have laws that are similar to the federal FOIA. A discussion of all such laws is beyond the scope of this chapter. However, California law provides one example.

California law declares that "access to information concerning the conduct of the people's business is a fundamental and necessary right of every person . . ." (California Government Code section 6250, 1995). Accordingly, the law further requires that "every person has a right to inspect any public record," during regular office hours, subject to certain enumerated restrictions (California Government Code section 6253(a), 2000). The law specifically directs enumerated government agencies to establish written guidelines and procedures for access to their records. Departments under such a requirement include the Department of Motor Vehicles, the Department of Youth Authority, the State Department of Health Services, and the Secretary of State (California Government Code section 6253.4(a), 2000). Specific records are exempted from disclosure including, but not limited to, personnel, medical or similar files pertaining to individuals; records of intelligence information of the Attorney General; records pertaining to pending litigation to which the public agency is a party; and interagency memoranda that are not retained by the public agency in the ordinary course of business (California Government Code section 6254, 2000). These provisions attempt to strike a balance between the public's right to know and the individual's interest in safeguarding his or her privacy and the confidentiality of information that pertains to him or to her.

Duty to Warn

Unlike many of the other limitations on confidentiality, which derive from state or federal statutes, the duty to warn is the product of case law, law that is set out by judges. The seminal case on duty to warn is *Tarasoff v. Regents of the University of California* (1976).

Tarasoff was a civil lawsuit brought by the parents of Tatiana Tarasoff against Poddar's former therapist and the therapist's employer, the University of California. Tatiana was killed by Prosenjit Poddar on October 27, 1969. Poddar had informed his psychologist two months prior to the killing that he intended to kill his former girlfriend. He did not name her, but she was easily identifiable, based on information given, as Tatiana. The therapist decided that Poddar should be committed for observation in a mental hospital. The therapist notified the campus police that he was going to request Poddar's commitment and asked the police to assist him in effectuating the confinement. The police took Poddar into custody, but then released him, believing that he would keep his promise to stay away from Tatiana. The director of the psychiatry department at the hospital then asked the police to return the therapist's letter, ordered that all notes regarding Poddar that had been taken by his therapist be destroyed, and instructed further that no action be taken to confine Poddar. Poddar persuaded Tatiana's brother to share a residence with him. Shortly after Tatiana returned home from a trip, Poddar went to her residence and killed her (*Tarasoff v. Regents of the University of California*, 1976).

Tatiana's parents sued for damages, claiming that (1) there was a duty to detain a dangerous patient, (2) there existed a duty to warn Tatiana or her parents that she was in grave danger, and (3) that the director of psychiatry had essentially abandoned a dangerous patient. Her parents alleged that the defendants had breached these duties and, as a result, Tatiana had suffered injury (death).

The defendants claimed that they could not have warned Tatiana, because to do so would have breached the confidentiality between the therapist and the patient that has traditionally been recognized and honored. The majority of the court, however, rejected this argument and held that when a patient "presents a serious danger . . . to another [person], [the therapist] incurs an obligation to use reasonable care to protect the intended victim against such danger" (*Tarasoff v. Regents of the University of California*, 1976: 340). That duty may be fulfilled by warning the intended victim, by notifying the police, or by taking any other steps that are reasonably necessary in view of the circumstances (*Tarasoff v. Regents of the University of California*, 1976: 340). The court specifically noted that the confidentiality afforded to therapists and their patients was not absolute:

> We recognize the public interest in supporting effective treatment of mental illness and in protecting the rights of patients to privacy and the consequent public importance of safeguarding the confidential character of psychotherapeutic communication. Against this interest, however, we must weigh the public interest in safety from violent assault....We conclude that the public policy favoring protection of the confidential character of patient-psychotherapist communications must yield to the extent to which disclosure is essential to avert danger to others. The protective privilege ends where the public peril begins. (*Tarasoff v. Regents of the University of California*, 1976: 346).

Many subsequent cases followed the reasoning of the court in *Tarasoff*. A New Jersey court ruled in *McIntosh v. Milano* (1979) that the doctor-patient privilege protecting confidentiality is not absolute, but is limited by the public interest or the private interest of the patient. In reaching this conclusion, the court relied on the 1953 case of *Earle v. Kuklo*, in which the court had found that a physician has a duty to warn third persons against possible exposure to contagious or infectious disease. Similarly, a Michigan appeals court held in *Davis v. Lhim* (1983) that a therapist has an obligation to use reasonable care whenever there is a person who is foreseeably endangered by his or her patient. The danger is foreseeable if the therapist knew or should have known, pursuant to his professional standard of care, of the potential harm.

Other courts have held similarly. In *Jablonski v. United States* (1983), a woman brought her violent boyfriend to the emergency room after he attempted to rape her mother. The psychiatrist concluded that, although the man was a danger to others, he could not be committed under California's involuntary commitment statute. The care provider did not request his past medical records, which indicated that he had been diagnosed with schizophrenia and had an extensive history of violent and threatening behavior. The care providers advised the girlfriend to stay away from him if she feared him. He later killed her. The court found that the hospital had failed to obtain important records and to adequately warn the victim.

Some courts have expanded the duty created by the *Tarasoff* court to unidentified victims. For instance, in *Lipari v. Sears, Roebuck & Co.* (1980), a patient attacked strangers in a nightclub with a shotgun that he had purchased from Sears. He had apparently not given his care providers at the Veterans Administration Hospital day care center where he was being treated any advance warning of his intended action. He did complain about his dissatisfaction with the care that he had been receiving. He terminated his psychiatric care approximately three weeks after purchasing the shotgun and shot up the nightclub approximately one month later, blinding one woman and killing her husband. The court rejected the argument that the duty enunciated in *Tarasoff* applied only to situations in which the victim was readily identifiable.

However, a number of courts have limited the duty imposed by *Tarasoff*. For instance, the courts in *Thompson v. Alameda County* (1980) and *Brady v. Hopper* (1983) found no duty to warn where there was no identifiable victim. The court in *Votteler v. Hartley* (1982) found no duty on the part of the therapist to warn the intended victim where the victim already had reason to know of the potential danger. In *Hosenei v. United States* (1982*)*, the court limited the duty to protect third parties to situations in which the therapist had the right to commit the patient to the hospital. In *Leonard v. Latrobe Area Hospital* (1993), the court found that there was no duty to warn the victim where the patient had not made specific threats against her. In a very recent case, the Supreme Court of Texas found that a health care provider does not have a duty to warn nonpatient third parties because there is no special relationship or connection between the health care provider and that person (*Van Horn v. Chambers*, 1998).

The extent to which the obligations imposed by *Tarasoff* and its progeny apply to therapists and other health professionals working in the capacity of researchers is unresolved. For instance, suppose that a licensed social worker or psychologist is interviewing individuals for a study that is investigating the association between substance use and violence. No counseling is provided as a part of the study, but individuals are provided with a referral list to social service organizations and self-help groups if they wish to have this information. One of the participants discloses to the psychologist conducting the interview that his girlfriend cheated him in a drug deal and he intends to "make her pay." The psychologist later learns from media reports that the man killed his girlfriend. The study had promised participants "confidentiality to the extent permitted by law." Numerous issues demand discussion here. Could the psychologist, working as a researcher, have foreseen that the participant would actually be dangerous, particularly in view of the lack of generally accepted standards by which to assess dangerousness (Lamb, Clark, Drumheller, Frizzell, and Surrey, 1989; Public Health Service, 1987)? Did the psychologist, acting in the capacity of a researcher and not a therapist, owe a legal duty to the girlfriend to warn her? If so, how is that to be balanced against his/her ethical and legal obligations to maintain the confidentiality of the information disclosed in the context of the research? As a matter of policy, should there be a privilege which excepts researcher-participant communications from disclosure in the context of civil litigation and criminal proceedings, as now exists between physicians and their patients and therapists and their patients? Could the psychologist, the research team, and/or the institution through which the research was being conducted be held responsible for having failed to disclose to the girlfriend the man's threat?

Suppose that instead of the study relating to substance use and violence, the study was designed to test the efficacy of a behavioral intervention to reduce risk behaviors for the transmission of HIV. A participant in the study finds that he has tested HIV-seropositive. He advises the licensed social worker who is conducting the baseline interviews on demographic characteristics and risk behaviors that he intends to take out as many people with him as he can, and the first person will be his girlfriend. Many questions must be answered to determine whether the social worker-interviewer is required by law to breach confidentiality and warn the girlfriend or take other measures to protect her. First, what is the state's interpretation/acceptance of the *Tarasoff* holding? The duty imposed by the *Tarasoff* case has not been accepted by the courts in all states. Second, how analogous is HIV infection to a shotgun wound? HIV is presently incurable, but not every unprotected sexual contact results in infection. Third, is the social worker working in the capacity of an interviewer and not a therapist bound by *Tarasoff*? And, finally, if the social worker is bound by *Tarasoff*, what steps must he or she take to satisfy the duty to warn?

Partner Notification Laws

"Partner notification" has been used most frequently to refer to the notification of an HIV-infected individual's current sexual or needle-sharing partner that he or she may have been exposed to HIV. Partner notification must be distinguished from contact tracing, which is a form of medical investigation, generally conducted by public health departments, that involves contacting all known sexual or needle sharing partners within a defined period of time to advise them of their possible exposure to HIV or to another sexually transmitted disease, and to ascertain their possible sexual and needle sharing partners who may have been exposed (Falk, 1988).

In some states, partner notification may be mandatory, while in others, it is voluntary. The requirement of partner notification demands analysis of the following questions: (1) Is this a disease for which partner notification is either mandatory or permissible? (2) Am I, as part of the research team, one of the individuals who is authorized or required under state law to notify someone's partner? California, for instance, previously had a voluntary notification procedure that applies to only physicians and surgeons. Consequently, individuals other than physicians or surgeons were neither required nor authorized to notify the partner of an HIV-infected individual of his or her possible exposure to HIV. And, California sets forth a specific procedure that was to be followed. First, the physician must try to obtain the patient's voluntary consent to notify the partner. If the patient refused, the physician was to notify the patient that he or she would notify the partner of the possible exposure. In notifying the partner, the physician could not reveal the identity or the identifying characteristics of the individual who may have exposed the partner to HIV. The physician was required to provide to the individual who he or she notified a referral for further counseling. The physician was not under an affirmative duty to notify the partner of the possible exposure. If the physician did notify the partner, the notification was to be done with the intent to interrupt the chain of transmission of HIV. Alternatively, the physician could have notified the county health officer and requested that the officer notify the partner of the HIV-infected patient (California Health and Safety Code section 199.25, 1990, repealed 1995).

Agency Audit Requirements

The sponsoring agency of the research may wish to review the records of a study and may, in providing the funding for the study, explicitly reserve the right to do so. For instance, as discussed above, the FDA may review records in connection with a clinical investigator-oriented inspection or a study-oriented inspection. In fact, regulations specifically provide that the FDA may have access to medical records (21 Code of Federal Regulations section 50.25, 1998). Although the FDA does not usually require the names of study participants, it may do so instances where a detailed examination of particular cases is warranted, or where there is reason to believe that the records do not represent actual cases or the actual results that were obtained (Food and Drug Administration, 1998). The agency, as well as the governing IRB, has an interest in ensuring compliance with the study protocol and the protections that were devised for the study participants.

Exercise

You are the principal investigator of an HIV prevention intervention trial for adolescents. Interviews with study participants include detailed information about sexual and substance using histories, as well as clinical information, such as history of sexually transmitted diseases.

1. What are the human subjects concerns that should be taken into account in conducting this study?

2. Explain how you will address each of the issues raised in (a) above.

3. Assume for the purpose of this question only that one of the study participants,

who is HIV infected, has informed you that he will not use safer sex practices. You know from the results of various tests that he is engaging in unprotected intercourse, and you believe, based on information from him, that he is having unprotected sexual relations with multiple partners. Discuss your ethical and legal obligations towards the study participant and towards his sexual partners.

4. Assume for the purpose of this question only that one of the adolescent girls participating in the study is pregnant. She is severely depressed and you believe that she may have had thoughts of suicide.

   a. Discuss all possible courses of action open to you.

   b. Identify the course of action that you believe is most appropriate in this situation and explain the ethical and legal basis for your decision.

Mechanisms to Enhance Confidentiality

Statutory and Regulatory Protections

There are various statutory provisions that, if utilized properly, can help to reduce the possibility that research data pertaining to a specific individual or individuals can be obtained through a subpoena.

The Privacy Act protects from disclosure records maintained by federal agencies which relate to individuals, absent the individual's written consent to access that information. The statute, however, permits disclosure in eleven circumstances:

1. to officers and employees of the agency maintaining the record, if they have the need for the record in the performance of their duties,
2. where disclosure is required by the Freedom of Information Act,
3. for a routine use, as defined by the statute,
4. to the Bureau of the Census for specifically enumerated purposes,
5. to individuals who have confirmed in writing that they would use the information as statistical research, and identifying information is excised from the record prior to its transfer,
6. to the National Archives of the United States, in various specified circumstances,
7. to another agency in the United States in connection with civil or criminal law enforcement activity,
8. "to a person pursuant to a showing of compelling circumstances affecting the health or safety of an individual if upon such disclosure notification is transmitted to the last known address of such individual,"
9. to a congressional committee or subcommittee of either house of Congress,
10. to the Comptroller General or his representatives, in the course of performing the duties of the General Accounting Office, and

11. pursuant to a court order (Privacy Act, 5 United States Code Annotated, section 552a(b), 1996, 1999).

The Public Health Services Act (1999) provides that:

> The Secretary may authorize persons engaged in biomedical, behavioral, clinical or other research (including research on mental health on the use and effect of alcohol and other psychoactive drugs) to protect the privacy of individuals who are the subject of research by withholding from all persons not connected with the conduct of such research the names or other identifying characteristics of such individuals. Persons so authorized to protect the privacy of such individuals may not be compelled in any Federal, State, or local, civil, criminal, administrative, legislative, or other proceedings to identify such individuals.

The scope of this protection is fairly broad in that it protects from disclosure both names and identifying characteristics, such as addresses and dates of birth, and covers proceedings at all levels of inquiry and in both the civil and criminal contexts. As such, it offers broader protection to the individuals participating in such research than do other statutes providing privacy protection from subpoenas and other legal proceedings.

Privacy protections are also available pursuant to statutory provisions relating to substance abuse and mental health research. The relevant statute provides that

> [r]ecords of the identity, diagnosis, prognosis or treatment of any patient which are maintained in connection with the performance of any program or activity related to substance abuse education, prevention, training, treatment, rehabilitation, or research, which is conducted, regulated or directly or indirectly assisted by any department or agency of the United State shall, except as provided in subsection (e), be confidential and be disclosed only [under specified circumstances]. (42 United States Code Annotated section 290dd-2, 1999)

Confidential information may be disclosed

> If authorized by an appropriate order of a court of competent jurisdiction granted after application showing good cause therefore, including the need to avert a substantial risk of death or serious bodily harm. In assessing good cause the court shall weigh the public interest and the need for disclosure against the injury to the patient, the physician-patient relationship, and to the treatment services. Upon the granting of such order, the court, in determining the extent to which any disclosure of all or any part of any record is necessary, shall impose appropriate safeguards against unauthorized disclosure. (42 United States Code Annotated

section 290dd—2(b)(2)(C), 1999)

A third statutory provision relates specifically to research relating to drug abuse and controlled substances. This provides that

> The Attorney General may authorize persons engaged in research to withhold the names and other identifying characteristics of persons who are the subjects of such research. Persons who obtain this authorization may not be compelled in any Federal, State, or local civil, criminal, administrative, legislative, or other proceeding to identify the subjects of research for which such authorization was obtained. (21 United States Code Annotated section 872(2), 1999)

The provisions of the above-mentioned statutes are not, however, self-enacting. That is, the researcher must apply for a confidentiality certificate in order to have the records protected under the relevant provision. This section sets forth as an example the procedures to be followed in applying for a certificate under the Public Health Services Act.

First, the researcher must verify that the research engaged in and for which he or she is seeking a certificate is research that is encompassed within the statutory provision: research on mental health, including the use and effect of alcohol and other drugs. Relevant regulations specify that the provisions for a confidentiality certificate are not applicable to research requiring an Investigational New Drug exemption or to approved new drugs or research related to law enforcement activities (42 Code of Federal Regulations section 2a.1(b), 1999).

An application for a confidentiality certificate must be submitted to the Office of the Director, National Institute on Drug Abuse, the Office of the Director, National Institute of Mental Health or the Office of the Director, National Institute on Alcohol Abuse and Alcoholism, 5600 Fishers Lane, Bethesda, Maryland 20857. The application may precede, accompany, or follow the submission of a grant to the Department of Health and Human Services. A separate application must be filed for each research project for which an authorization of confidentiality is being requested (42 Code of Federal Regulations section 2a.3, 1999).

The application must contain the following information: the name and address of the investigator responsible for the research, the name of the sponsor or institution with which the researcher is affiliated, the location of the research project, a description of the research facilities, the names and addresses of all personnel having major responsibility for the research, a description of the personnel's experience, an outline of the research protocol, the date on which the research will begin or has begun and the projected date of completion, a specific request signed by the individual responsible for the research for authority to withhold the names and other identifying information relating to the research participants, the reasons for the request of authority, and assurances that the persons applying for the certificate will apply with federal regulation relating to the protection of human subjects, that a grant of authority will not be represented as an endorsement of the research, and that specified information will be provided to the research participants (42 Code of Federal Regulations section 2a.4, 1999). If granted, the certificate is not transferable and applies only to the names and other identifying characteristics of the individuals participating in the single research project specified in the confidentiality certificate (42 Code of Federal Regulations section 2a.6, 1999).

Other Legal Mechanisms

A subpoena can be challenged in state or federal court, depending on the court that issued the subpoena. The procedures to do so vary from state to state. For this reason, this discussion focuses only on procedures in the federal courts. Only a brief summary of the procedures are provided here. Any investigator served with a subpoena should consult with the attorney for his or her institution or company.

A subpoena can be challenged through a mechanism known as a motion to quash a subpoena. This must be made promptly after the subpoena is issued. The motion must be made by the person from whom the things or records are being requested. Generally, a person who is not the person or entity from whom the records are being sought will have no right to bring a motion to quash a subpoena or a motion for a protective order (*Vogue Instrument Corporation v. Len Instruments Corporation*, 1967). The only exception is where a person claims some right or personal privilege with respect to the documents or records that are being sought (*Norris Manufacturing Company v. R.E. Darling Company*, 1976).

The court may decide to quash or modify the subpoena if it finds that it is unreasonable or oppressive (Federal Rules of Civil Procedure 45(b), 1983). A researcher could make various arguments with respect to the unreasonableness or oppressiveness of a subpoena. For instance, the disclosure of the information requested could result in a disruption of the research and a lack of confidence in both the researcher and the study. This argument can be supported with affidavits from other researchers, physicians, or agencies who refer individuals to participate in the study. For instance, social service agencies and clinics helping to recruit individuals to the study may refuse to continue in this role if confidentiality will be breached.

The researcher could also indicate that the search for the records that are requested would consume an inordinate amount of time and an excessive dedication of resources. This could be true, for instance, where the data base for a particular study is very large and extensive resources would be required to pull out from those data the information requested by the subpoena. This argument may not be particularly successful, because the court has the power to deny the motion to quash the subpoena conditionally on the advancement by the person or entity seeking the records of the reasonable costs of producing the documents or materials requested (Federal Rule of Civil Procedure 45(b), 1983).

A third argument that can be made is that the information being sought is not relevant to the litigation. It is difficult to make this argument successfully, because the discovery process extends to information that is not privileged (specifically protected under certain recognized exceptions), and that is relevant to the subject matter of the pending lawsuit (Federal Rule of Civil Procedure 26(b), 1983). As an example, assume that an individual participating in an intervention trial of a new behavioral approach to alcohol abuse has a car accident. The insurance company wants to access his research records, claiming that these are relevant to his drinking behaviors. The researcher could try to argue that the individual's behavior was not relevant to his behavior at the time of the accident.

Melton and Gray (1988) have argued that a common-law privilege protecting researcher-participant communication should be recognized. This would be critical, for instance, in resolving *Tarasoff*-type situations involving researchers. One of the reasons that the court had focused so heavily on confidentiality in the *Tarasoff* case was because of the existence of a long-recognized privilege excepting from discovery in most circumstances communications between physicians and their patients and therapists and their patients. Most jurisdictions do not have such a privilege for researchers and their research participants. In order to demonstrate that such a privilege should be recognized, the communication must have taken place in the context of a confidential relationship, the relationship must be important to the community, and the injury that would result from a breach in confidentiality would be greater than the benefit that would be gained.

As an example, consider the many prosecutions that have occurred recently against women who have ingested controlled substances during their pregnancies (Chambers, 1986; Churchville, 1988; *Commonwealth v. Kemp*, 1994; *Commonwealth v. Pelligrini*, 1990; Hager, 1992; Jacobus, 1992; *Johnson v. State*, 1992; *Kentucky v. Welch*, 1993; *People v. Hardy*, 1991; *People v. Morabito*, 1992; *Reinesto v. Superior Court*, 1995; Rubin 1990; *Sheriff of Washoe County v. Encoe*, 1994; *State v. Gray*, 1992; *State v. Kruzicki*, 1995; *State v. Luster*, 1992; *Whither v. State*, 1996). Whether or not one agrees with prosecution as an appropriate response to maternal use of controlled substances during pregnancy, consider the potential consequences if research participants' records could be accessed and utilized as the basis of prosecution, or in furtherance of prosecution, against them. Such access could have the potential to discourage any participation and could result in an inability to better understand substance use and effective strategies for intervention.

CHAPTER SUMMARY

This chapter addressed the administrative consequences that could result from a breach of an ethical responsibility to the participants in research or to the institution. It should be noted again that ethical obligations are not invariably coextensive with legal obligations and vice versa. As we saw in chapter 4, the overriding theme is the protection of the research participants and the consequences of a failure to do so. However, it is also clear from the discussion in chapter 5 that there are legally-imposed limits on the extent to which participants may be protected from the disclosure of information about them and/or their behavior.

References

Alaska Statutes § 47.24.010 (1998).
Andrews v. Eli Lilly & Company, 97 F.R.D. 494 (N.D. Ill. 1983).
Application of American Tobacco Company, 880 F.2d 1520 (2d Cir. 1989).
Application of R.J. Reynolds Tobacco Company, 136 Misc.2d 282, 328 N.Y.S. 729 (1987).

Baringa, M. (1992). Who controls a researcher's files? *Science, 256*, 1620-1621.
Brady v. Hopper, 570 F. Supp. 1333 (D. Colo. 1983).
Broad, W. & Wade, N. (1982). *Betrayers of the Truth*. New York: Simon and Schuster.
California Government Code §§ 6250, 6253, 6254 (West 1995).
Chambers, M. (1986, October 9). Dead Baby's Mother Faces Criminal Charge on Acts in Pregnancy, *N.Y. Times*.
Churchville, V. (1988, July 23). D.C. Judge Jails Woman as Protection for Fetus, *Washington Post*, at A1.
Cobbs v. Grant, 8 Cal. 3d 229 (1972).
Commission on Research Integrity, United States Department of Health and Human Services. (1998). Integrity and Misconduct in Research.
Commonwealth v. Kemp, 643 A.2d 705 (Pa. 1994).
Commonwealth v. Aurick, 19 A.2d 920 (Pa. 1941).
Commonwealth v. Malone, 47 A.2d 445 (Pa. 1946).
Commonwealth v. Pellegrini, No. 87970, slip. op. (Mass. Dist. Ct. Oct. 15, 1990).
Commonwealth v. Welansky, 5 N.E.2d 902 (Mass. 1944).
Connecticut General Statutes Annotated § 17b-450 (West 1998).
Council for International Organizations of Medical Sciences (CIOMS), World Health Organization (WHO). (1993). *International Guidelines for Biomedical Research Involving Human Subjects*. Geneva: WHO.
Council for International Organizations of Medical Sciences (CIOMS). (1991). *International Guidelines for Ethical Review of Epidemiological Studies*. Geneva: Author.
Davis v. Lhim, 124 Mich. App. 291 (1983), *affirmed on remand* 147 Mich. App. 8 (1985), reversed on grounds of government immunity in *Canon v. Thumudo*, 430 Mich. 326 (1988).
Deitchman v. E.R. Squibb & Sons, Inc., 740 F.2d 556 (7th Cir. 1984).
Department of Justice v. Tax Analysts, 109 S.Ct. 2841 (1989).
Departmental Appeals Board (DAB) Decision No. 1677 (February 5, 1999).
Director, OPRR. (1997, January 23). Memorandum to OPRR Staff re: Compliance Oversight Procedures. Available at http://grants.nih.gov/grants/oprr/oprrcomp.htm.
Dresser, R. (1993a). Defining scientific misconduct: The relevance of mental state. *Journal of the American Medical Association, 269*, 895-897.
Dresser, R. (1993b). Sanctions for research misconduct: A legal perspective. *Academic Medicine, 68*, S39-S43.
Earle v. Kuklo, 26 N.J. Super. 471 (App. Div. 1953).
Farnsworth v. Proctor and Gamble Company, 785 F.2d 1545 (11th Cir. 1985).
Farthing, M.J.G. (1998). An editor's response to fraudsters. *British Medical Journal, 316*, 1729-1731.
Federal Rules of Civil Procedure 26(b), 45(b) (1983).
Florida Statute Annotated § 415.102(12) (West 1999).
Florida Statute Annotated § 415.102(22) (West 1999).
Food and Drug Administration (FDA). (1998, September). Information Sheets: Guidance for Institutional Review Boards and Clinical Investigators. 1998 Update. Available at http://www.fda.gov/oc/oha/IRB/toc9.html.
Forsham v. Harris, 445 U.S. 169 (1980).
Freedman, B. & Glass, K.C. (1990). *Weiss v. Solomon*: A case study in institutional responsibility for clinical research. *Law, Medicine & Health Care, 18*, 395-403.
Friter v. Iolab Corporation, 607 A.2d 1111 (Pa. Super. Ct. 1992).
Gore, A., Jr. (1981). Testimony before the House Subcommittee on Investigations and Oversight of the Committee on Science and Technology. *Fraud in Medical Research: Hearings Before the House Subcommittee on the Investigation and Oversight of the Committee on Science and Technology*. 97th Congress, 1st Session, 286.
Gray v. Grunnagle, 223 A.2d 663 (Pa. 1966).
Hager, P. (1992, Aug. 22). Murder Charge Rejected in Drug-Related Stillbirth, *Los Angeles Times*, at A1, col. 5.
Hilts, P.J. (1991, March 22). Hero in exposing science hoax paid dearly. *New York Times*, A1, B6.
Hilts, P.J. & Stolberg, S.G. (1999, May 13). Ethics lapses at Duke halt dozens of human experiments. *New York Times*, A26.
Hiserodt v. Shalala, (1994, July 20). C.A. No. 91-0224 (W.D. Pa.), *affirmed* No. 94-3404 (3d Cir. July 5, 1995). See http://ori.dhhs.gov/legal/htm.
Holder, A.R. (1989). Researchers and subpoenas: The troubling precedent of the Selikoff case. *IRB, 11*, 8-11.
Holder, A.R. (1993). Research records and subpoenas: A continuing issue. *IRB, 15*, 6-7.

Horne v. Patton, 287 So.2d 824 (Ala. 1974).
Horowitz, A.M. (1996). Fraud and scientific misconduct in the United States. In S. Lock & F. Wells (Eds.). *Fraud and Misconduct in Medical Research* (2nd ed., pp. 144-165). London: BMJ Publishing Group.
Hoseini v. United States, 541 F. Supp. 999 (D. Md. 1982).
In re Grand Jury Subpoena Dated Jan. 4, 1984, 750 F.2d 223 (2d Cir. 1984).
Jablonski v. United States, 712 F.2d 391 (9th Cir. 1983).
Jacobus, P. (1992, April 2). Prosecutors' New Drug War Target: The Womb, *Los Angeles Daily Journal* at 1, col. 2.
John C. Hiserodt, M.D., Ph.D. (1995, February 25). DAB No. 1466.
Johnson v. State, 602 So. 2d 1288 (Fla. 1992).
Kentucky v. Welch, 864 S.W.2d 280 (Ky. 1993).
Kissinger v. Reporters Committee for Freedom of the Press, 445 U.S. 136 (1980).
Kubie, L.S. (1958). *Neurotic Distortion of the Creative Process*. Lawrence, Kansas: University of Kansas Press.
Kus v. Sherman Hospital, 644 N.E.2d 1214 (Ill. App. 1995).
Kuzma, S.M. (1992). Criminal liability for misconduct in scientific research. *University of Michigan Journal of Law Reform, 25*, 357-421.
Lamb, D.H., Clark, C., Drumheller, P., Frizzell, K., & Surrey, L. (1989). Applying *Tarasoff* to AIDS related psychotherapy issues. *Professional Psychology: Research and Practice, 20*, 37-43.
Leonard v. Latrobe Area Hospital, 625 A.2d 1228 (Pa. Super. Ct. 1993).
Lipari v. Sear, Roebuck & Co., 497 F. Supp. 185 (D. Neb. 1980).
Lock, S. (1996). Research misconduct: A resume of recent events. . In S. Lock & F. Wells (Eds.). *Fraud and Misconduct in Medical Research* (2nd ed., pp. 14-39). London: BMJ Publishing Group.
McBride, G. (1974). The Sloan-Kettering Affair: Could it have happened anywhere? *Journal of the American Medical Association, 229*, 1391-1410.
McIntosh v. Milano, 168 N.J. Super. 466 (1979).
McTaggart, L. (1980, December 7). Putting drug testers to the test. *New York Times*, Magazine Section, 174.
Melton, G.B. & Gray, J.N. (1988). Ethical dilemmas in AIDS research: Individual privacy and public health. *American Psychologist, 43*, 60-64.
Mink v. University of Chicago, 460 F. Sup. 713 (N.D. Ill. 1978).
Mints, M. (1979, June 1). Indictment accuses drug-testing firm of falsifying results. *Washington Post*, A9.
Moore v. Regents of the University of California, 215 Cal. App. 3d 709 (1988), *reversed* 51 Cal. 3d 120 (1990).
Needleman v. Shalala, et al. (1996, May 22). C.A. No. 920749 (W.D. Pa.). See http://ori.dhhs.gov/legal/htm.
Nevada Revised Statute § 200.5092 (1987).
Nevada Revised Statute § 200.5093 (1987).
New York Penal Law § 125.25 (19__).
Nigg, N.G. & Radulsecu, G. Scientific misconduct in environmental science and toxicology. *Journal of the American Medical Association, 272*, 168-170.
Nylenna, M., Aderse, D., Dahlquist, G., Sarvas, M., & Aakvaag, A. (1999). Handling of scientific dishonesty in the Nordic countries. *Lancet, 354*, 57-61.
Office for Protection from Research Risks Review Panel. (199, June 3). Report to the Advisory Committee to the Director, NIH. Available at http://www.nih.gov/welcome/director/060399b.htm.
Office of Research Integrity. (1996). Consequences of being accused of scientific misconduct. In 1996 ORI Annual Report.
Office of Research Integrity. (1995). Annual Report.
Office of Research Integrity. (1996). Annual Report.
Office of Research Integrity. (1994a). Chagnon. 59 Federal Register 38979.
Office of Research Integrity. (1994b). Constantoulakis, 59 Federal Register 45670.
Office of Research Integrity. (1994c). Leisman. 59 Federal Register 64667.
Office of Research Integrity. (1995). 60 Federal Register 66276.
Office of Research Integrity. Model Procedures for Responding to Allegations of Scientific Misconduct. Available at http://ori.dhhs.gov/.
Office of Research Integrity. Summaries of Closed Investigations Not Resulting in Findings of Misconduct. Available at http://ori.dhhs.gov/.
Office of Research Integrity. (1994d). Suprenant. 59 Federal Register 39366.

Office of Scientific Integrity. (1999, January). Scientific Misconduct Investigations, 1993-1997. Rockville, Maryland: Department of Health and Human Services.
Ohio Revised Code Annotated § 5101.60(K) (Banks-Baldwin 1994).
Ohio Revised Code Annotated § 5101.61(A) (Banks-Baldwin 1994).
ORI Newsletter (1994, December). Available at http://ori.dhhs.gov/definition.htm.
Palca, J. (1992). The case of the Florida dentist. *Science, 255,* 392-394.
Parrish, D.M. (1997). Improving the scientific misconduct hearing process. *Journal of the American Medical Association, 277,* 1315-1319.
Pennsylvania Statute Annotated, title 35, § 10225.103 (1999).
People v. Hardy, 469 N.W.2d 50 (Mich. App. 1991).
People v. Morabito, (City Ct., Geneva), *reported in* N.Y.L.J. Feb. 19, 1992, at 30
Public Health Service, United States Department of Health and Human Services. (1987*). A Public Health Challenge: State Issues, Policies, and Programs.* Vol. 1, pp. 4-1 to 4-31.
Reinesto v. Superior Court, 894 P.2d 733 (Ariz. App. 1995).
Riis, P. (1006). Creating a national control system on scientific dishonesty within the health sciences. . In S. Lock & F. Wells (Eds.). *Fraud and Misconduct in Medical Research* (2nd ed., pp. 114-127). London: BMJ Publishing Group.
Riis, P. (1994). Prevention and management of fraud—in theory. *Journal of Internal Medicine, 235,* 107-113.
Rubin, N. (1990, June/July). Motherhood on trial, *Parenting,* 74.
Ruling on Respondent's Motion to Dismiss for Lack of Jurisdiction: Notice of Further Procedures,. (1995, Aug. 15). Board Docket No. A-95-123 (Kerr). See http://ori.dhhs.gov/legal.htm.
Shapiro, M.F. & Charrow, R.D. (1985). Special report: Scientific misconduct in investigational drug trials. *New England Journal of Medicine, 312,* 731-___.
Sheriff of Washoe County v. Encoe, 885 P.2d 596 (Nev. 1994).
State v. Gray, 584 N.E.2d 710 (Ohio 1992).
State v. Hefner, 155 S.E. 879 (N.C. 1930).
State v. Kruzicki, 541 N.W.2d 482 (Wis. App. 1995);
State v. Luster, 419 S.E.2d 32 (Ga. App. 1992).
Tarasoff v. Regents of the University of California, 551 P.2d 334 (1976).
Texas Human Resources § 48-002 (West Supp. 1999).
Thompson v. County of Alameda, 27 Cal. 3d 741 (1980).
United States v. Breuning, No. K-88-0135 (D. Md. Sept. 18, 1988). Cited in S.M. Kuzma. (1992).Criminal liability for misconduct in scientific research. *University of Michigan Journal of Law Reform, 25,* 357-385.
United States Department of Health and Human Services. (1993, September). Office of Research Integrity: An Introduction. Rockville, Maryland: United States Department of Health and Human Services.
United States Department of Health and Human Services. (1999, October 22). Promoting integrity in research. HHS Fact Sheet. Available at http://ori.dhhs.gov/991022a.htm.
Van Horn v. Chambers, 970 S.W.2d 542 (Tex. 1998).
Votteler v. Heltsley, 327 N.W.2d 759 (Iowa 1982).
Washington Research Project v. Department of Health, Education & Welfare, 504 F.2d 238 (D.C. Cir. 1974).
Wenger, N.S., Korenman, S.G., Berk, R., & Berry, S. (1997). The ethics of scientific research: An analysis of focus groups of scientists and institutional representatives. *Journal of Investigative Medicine, 45,* 371-380.
Whither v. State, 65 U.S.L.W. 2066 (S.C. 1996).
Wion, A.H. (1979). The definition of "agency records" under the Freedom of Information Act. *Stanford Law Review, 31,* 1093-1115.
21 Code of Federal Regulations §§ 50.3, 50.25 (1998).
42 Code of Federal Regulations §§ 2a.1-2a.7, 50.102, 50.104 (1999).
45 Code of Federal Regulations §§ 76.110, -.115, -.305, -.311, -.313-.314, -.320, -.400-.413, -.415 (1999).
45 Code of Federal Regulations § 689.1 (1998).
5 United States Code Annotated §§ 552, 552a (West 1996 & Supp. 1999).
5 United States Code § 553 (West Supp. 1996).
18 United States Code §§ 287, 1001, 1341 (West 1999)
18 United States Code § 1505 (West 1998).
21 United States Code Annotated § 872 (West 1999).
42 United States Code Annotated §§ 241, 290dd-2 (West 1999).

APPENDIX 1

PRINCIPLES OF RESEARCH DESIGN

Each discipline that concerns itself with health research--sociology, anthropology, epidemiology, psychology, and health promotion, for example--maintains a lexicon specific to that field that constitutes, in effect, a shorthand for describing the study designs, relevant measures and measurements, and interpretational issues that constitute the core of that discipline's research methodology. It is, consequently, impossible to address specifically the concerns across all relevant disciplines or to utilize each discipline's language in discussing issues pertaining to study design. Accordingly, study design and issues affecting the interpretation of one's results are discussed here using epidemiology as a framework, with the understanding that many of the broader concepts, if not the language itself, are applicable across disciplines. Because relevant measures and measurements differ so greatly across fields, they are not discussed here. Rather, readers are urged to consult references in their specific disciplines.

CAUSATION AND CAUSAL INFERENCE

Epidemiology seeks to answer such questions as "Why do some people contract certain illnesses more frequently than others?" and "Why does a specific illness progress more rapidly in some people compared to others?" The simplest approach to causality in such circumstances is that of pure determinism.

Pure determinism posits specificity of cause and specificity of effect, *i.e.,* that the factor being examined as a cause is the one and only cause of the disease under examination, and that the disease under investigation is the only effect of that factor (Kleinbaum, Kupper, & Morgenstern, 1982). This implies that the factor under examination is both a necessary and sufficient cause of that disease.

Robert Koch's formulation of the criteria for disease causality, in essence, operationalized pure determinism (Kleinbaum, Kupper, & Morgenstern, 1982). Koch's work on tuberculosis provided the basis for his refinements of causation criteria to include five elements:

T1. An alien structure must be exhibited in all cases of the disease.
T2. The structure must be shown to be a living organism and must be distinguishable from all other micro-organisms.
T3. The distinction of micro-organisms must correlate with and explain the disease phenomena.
T4. The micro-organism must be cultivated outside the diseased animal and isolated from all disease products which could be causally significant.

> T5. The pure isolated micro-organism must be inoculated into test animals and these animals must then display the same symptoms as the original diseased animal (Carter, 1985).

Koch's causational model has been criticized for its various limitations, including its failure to recognize (1) the multifactorial etiology of many diseases; (2) the multiplicity of effects associated with specific factors; (3) the complexity of many causal factors; (4) our incomplete understanding of disease and disease processes; and (5) the limitations inherent in our ability to measure the causal process (Kleinbaum, Kupper, & Morgenstern, 1982).

Modified determinism addresses the limitations of Koch's model. Rothman explains this model as follows:

> A *cause* is an act or event or a state of nature which initiates or permits, alone or in conjunction with other causes, a sequence of events resulting in an *effect*. A cause which inevitably produces the effect is *sufficient*....
>
> * * * *
>
> A specific effect may result from a variety of different sufficient causes....If there exists a component cause which is a member of every sufficient cause, such a component is termed a *necessary cause*....(Rothman, 1976).

Hence, this model recognizes both that a cluster of factors, rather than a single agent, may produce an effect, and that a specific effect may be the product of various causes. The strength of a specific causal factor depends upon the relative prevalence of component causes. A factor, even though rare, may constitute a strong cause if its complementary causes are common (Rothman, 1986). Two component causes of a sufficient cause are said to be synergistic, in that their joint effect exceeds the sum of their separate effects (Rothman, 1976). As an example, individuals exposed to asbestos are at increased risk of cancer if they also smoke (Hammond, Selikoff, & Seidman, 1979).

This modified deterministic model, however, also has its limitations. We are often unable to identify all of the components of a sufficient cause (Rothman, 1986). Consequently, epidemiologists utilize probability theory and statistical techniques to assess the risk of disease resulting from exposure to hypothesized causal factors (Rothman, 1986). Causation may be inferred by formulating general theories from observation (induction) or by testing general theories against observation (deduction)(Weed, 1986).

Using an inductive approach, Hill has enunciated the criteria to be considered in identifying causal associations: (1) strength; (2) consistency; (3) specificity; (4) temporality; (5) biological gradient; (6) plausibility; (7) coherence; (8) experimental evidence; and (9) analogy (Morabia, 1991). Each of these criteria is discussed briefly below.

The "strength" of an association between the putative causal factor and the effect is dependent on the relative prevalence of other component causes. This criterion encompasses two separate issues: the frequency with which the factor under investigation is found in cases of a specific disease and the frequency with which the factor occurs in the absence of the disease (Sartwell, 1960). "Consistency" refers to the repeated observation of an association between the putative causal factor and the effect in varied populations at varied points in time and under different circumstances (Susser, 1991). Inconsistency, however, does not necessarily negate a causal relationship because all causal components must exist to bring about the effect, and some may be absent. "Specificity" refers to the association between a postulated cause and a single effect (Rothman, 1986). Hill specifically cautioned against overemphasizing the importance of this particular element (Hill, 1965).

"Temporality" requires that the cause precede the effect in time (Rothman, 1986). Although a dose-response curve, or "biological gradient," is to be considered, it does not necessarily indicate causation due to the effects of confounders. "Plausibility" requires that the hypothesized relationship between the causal factor and the effect be biologically plausible. This is clearly limited by the state of our knowledge at any point in time. "Coherence" requires that a postulated causal association be consistent with our knowledge of the natural history and biology of the disease in question. Experimental evidence is rarely available for human populations. "Analogy" posits that reference to known examples, such as a causal association between one drug and birth defects, may provide insights into other causes of birth defects, such as another drug (Rothman, 1986).

RESEARCH DESIGN

Research design provides a means by which to examine the relation of cause and effect in a specific population (Susser, 1991). This section explores the basic study designs used in epidemiological research, as well as several hybrid designs. Other resources should be consulted for a more extensive discussion.

Types of Research

Epidemiologic research can be classified into three major types: experimental, quasi-experimental, and observational. With each type of research, a variety of different designs can be utilized. Each of these types is described briefly below, followed by a discussion of specific study designs.

Experimental Research

Experimental research involves the randomization of individuals into treatment groups, also known as study arms, *i.e.*, individuals are randomly allocated to receive a particular treatment under investigation (treatment group) or an alternative treatment or placebo (control group). Randomization is considered essential as a safeguard against selection bias and as insurance against accidental bias (Gore, 1981), *i.e.*, to ensure that both groups are representative of the population as a whole or that both groups are similar in all respects except the treatment that is under study. Experimental studies conducted in the laboratory are usually of relatively short duration and are often used to test etiologic hypotheses, to estimate treatment effects, or to examine the efficacy of an intervention (Kleinbaum, Kupper, & Morgenstern, 1982).

Clinical trials are experimental studies of much longer duration. They have been referred to as "the epidemiologic 'gold standard' for causality inference" (Gray-Donald & Kramer, 1988: 885). They are usually conducted to test the efficacy of a specific intervention, to test etiologic hypotheses, or to estimate long term health effects. Both the experimental group and the control group consist of patients who have already been diagnosed with the disease of interest or are at risk of the disease of interest in a clinical trial involving a prevention. Additionally, the patients must have agreed to be randomized to one of the study arms. Clinical trials often utilize double blinding, whereby neither the individuals conducting the study nor the participants in the study know who is in the treatment group and who is in the control group (Senn, 1991).

Community interventions are also usually of longer duration. They are often initiated to test the efficacy and the effectiveness of a particular health intervention.

Experimental studies offer numerous advantages, including the ability, through randomization, to control for extraneous factors that may be related to the outcome under examination. Unfortunately, the study population ultimately selected through this process may not be comparable to the target population with respect to important characteristics.

Quasi-Experiments

Quasi-experiments involve the comparison of one group to itself or of multiple groups. Although this study design permits the investigator to manipulate the study factor, as in an experiment, randomization is not used. Quasi-experiments are most often conducted in a clinic or laboratory setting to test etiologic hypotheses, to evaluate the efficacy of an intervention, or to estimate the long term health effects of an intervention. Those conducted in the program and policy arenas are often devoted to the evaluation of programs or interventions or to an analysis of the costs and benefits of an intervention (Kleinbaum, Kupper, & Morgenstern, 1982). Quasi-experiments are generally smaller and less expensive than experimental studies. However, the investigator has less control over the influence of extraneous risk factors due to the lack of randomization (Kleinbaum, Kupper, & Morgenstern, 1982).

Observational Studies

Observational studies are the most frequently utilized type of study in epidemiology. Unlike experimental and quasi-experimental studies, they do not involve the manipulation of the study factor. The goal of observational studies is to arrive at the same conclusions that would have resulted from an experiment (Gray-Donald & Kramer, 1988). Observational studies may be descriptive or etiologic in nature. Descriptive studies are often used to estimate the frequency of a specific disease in a population or to generate hypotheses or ideas for new interventions. Etiologic studies can be used not only to generate etiologic and preventive hypotheses, but can also be used to test specific hypotheses and estimate health effects. Observational studies afford the investigator much less control over extraneous risk factors than do either experimental or quasi-experimental studies because they are conducted in a natural setting (Kleinbaum, Kupper, & Morgenstern, 1982).

Observational studies can use any of numerous study designs, depending on how the research question is framed, the current state of knowledge with respect to the disease or exposure/risk factor at issue, the costs of the proposed study, and various other considerations. Observational study designs include cohort study design, case-control design, and cross-sectional design. Other common observational designs include the etiologic study, the proportional study, space-time cluster studies, and the family cluster study (Kleinbaum, Kupper, & Morgenstern, 1982).

Clinical trials and observational studies are particularly central to epidemiology. For this reason, clinical trials and various types of observational study designs are addressed in further detail below.

Clinical Trials

Clinical trials of new drugs are conducted in three phases. The first phase consists of the initial introduction of the drug into humans. Phase I studies are conducted to assess the metabolic and pharmacologic actions of the drugs in humans, the side effects of the drug, and the effectiveness of the drug. Phase I studies, which are closely monitored, are generally limited to 20 to 80 participants (21 Code of Federal Regulations section 312.21(a), 1998).

Phase II studies build on the knowledge that was obtained from Phase I studies. Phase II studies are carried out in patients with the disease under study for the purpose of assessing the drug's effectiveness for a particular indication and for determining the drug's short-term side effects and risks. Typically, Phase II studies involve up to several hundred participants (21 Code of Federal Regulations section 312.21b, 1998).

Phase III studies incorporate the knowledge gained through Phase I and Phase II trials. Phase III trials focus on gathering additional data relating to the drug's effectiveness and safety. Phase III studies are potentially quite large, sometimes involving thousands of participants (21 Code of Federal Regulations section 312.21c, 1998).

The initial phase of planning a clinical trial requires that a decision be made regarding the treatment(s) to be studied and the eligibility criteria for individuals to participate in the trial (Rosner, 1987). Clinical trials designed to answer questions relating to biological response generally enroll a rather homogenous participant population, in order to reduce the variability between participants and simplify the analysis of the results. Patients enrolled in such a study must be sufficiently healthy so that they do not die before the end of the study, but not so well that they recover from the disease.

Larger clinical trials, particularly multisite trials, often enroll more heterogeneous participants. This situation more closely mirrors everyday clinical practice with diverse patients and also comports with ethical considerations and federal regulations regarding the equitable distribution of the benefits and the burdens of research.

During a phase III clinical trial, the treatment under study is compared tone or more standard treatments, if a treatment exists, or to a placebo. A placebo has been defined as

> Any therapy (or that component of any therapy that is intentionally or knowingly used for its nonspecific, psychological, or psychophysiological therapeutic effect, or that is used for a presumed specific therapeutic effect on a patient, symptom, or illness but is without specific activity for the condition being treated...[and], when used as a control in experimental studies, is a substance or procedure that is without specific activity for the condition being treated. (Shapiro and Shapiro, 1997: 41)

Participants are randomized to one of the treatment arms or, if a placebo arm is used, to the treatment arm(s) or placebo arm. The "gold standard" for clinical trials requires that neither the researcher(s) nor the participants know until the conclusion of the study which individuals are receiving the experimental treatment and which are receiving the standard treatment or placebo, i.e. the study is double-blinded.

Additionally, researchers must decide which endpoint(s) to use as a basis for evaluating participants' response to the treatment (O'Brien & Shampo, 1988). The endpoint(s) will vary depending on the disease and the treatment under investigation, but may include recurrence of an event, such as myocardial infarction; quality of life; functional capacity; or death. Clinical trials are often concerned not only with whether an endpoint occurs, but also with the length of time until its occurrence. As an example, a clinical trial for a cancer treatment may be concerned with not only the recurrence of the malignancy, but also with the time between the treatment and the reappearance of the disease.

Ethical concerns have been raised about classic clinical trials on a number of grounds. Clinicians may believe that patients should not be randomized because it may deprive them of an opportunity to receive a new alternative drug (Farrar, 1991). Others believe that randomization is inappropriate in situations in which the patient has exhausted all available therapies or one in which the standard therapy has provided no benefit (Rosner, 1987).

Crossover designs have also been proposed as an alternative to classic clinical trials and as a mechanism for addressing variations between patients in response to treatment. With a crossover design, half of the patients are randomized to Group 1 and the second half to Group 2. Following administration of Treatment A to Group 1 and Treatment B to Group 2, the allocation of treatment is reversed. Generally, crossover designs incorporate an appropriate "washout period" between administration of the treatment to each group, in order to reduce the possibility of a carryover effect from the first treatment period to the second (Hills & Armitage, 1979). Crossover designs are most useful in situations in which the treatment under investigation is for the alleviation of a condition, rather than the effectuation of a cure.

Observational Designs

Cohort Studies

Cohort studies can be conducted prospectively or retrospectively. Prospective cohort studies require the identification and classification of initially disease-free individuals into categories according to whether they have or have not been exposed to the factor under study. Each group is followed over time in order to observe the number of new cases of the disease under investigation that occurs in each group in a specified period of time (Kelsey, Thompson, & Evans, 1986).

Numerous difficulties may attend prospective studies. First, individuals may already have been exposed to the factor under study and the length and intensity of that exposure may be difficult to ascertain. As an example, a cohort study to examine the effects of an occupational exposure would consist of a group exposed to the substance under study and a group that was not exposed. Depending upon the particular industry and configuration of the workplaces, however, some members of the group classified as unexposed may, in fact, have been exposed to small amounts of the substance. Second, although individuals enrolled into the study may be believed to be free of the disease, the disease process may, in fact, have commenced in some but may be undetectable by diagnostic methods and tools then available. This could be true, for instance, in studies involving cancer or schizophrenia. Third, a minimum length of time following exposure may be required to allow a biologically appropriate induction time, as well as a subsequent period of time after causation but before disease detection (latent period). In situations where we do not have complete knowledge of the induction and latency periods, we must make assumptions about the lengths of these times (Rothman, 1986). Fourth, prospective cohorts also require large sample sizes and are often quite costly (Kelsey, Thompson, & Evans, 1986). (What is considered a "large sample size" varies depending upon the disease under study, the exposure under study, and various other factors.) Fifth, the choice of a comparison group of unexposed individuals may be quite difficult. Too, we tend to think of disease as being present or absent, but some diseases, such as high blood pressure, may occur along a spectrum, making classification of individuals as diseased or not diseased more complex (Kelsey, Thompson, & Evans, 1986).

Individuals in a prospective cohort study are to be followed over time, as is indicated in Figure 1 below. However, some individuals may drop out of the study or be lost to follow-up. These losses may be related to disease status, thereby producing a bias in the measurement of disease (Kelsey, Thompson, & Evans, 1986). It is easy to imagine that as someone becomes progressively more ill, that he or she may not want to undergo a physical examination or respond to questions relating to the illness. Information on other extraneous variables that may affect the results may not be available (Kelsey, Thompson, & Evans, 1986).

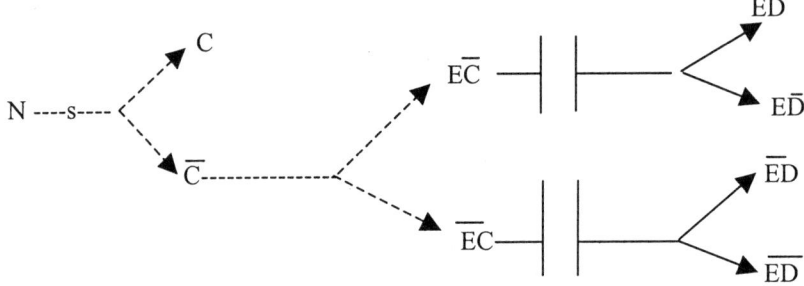

N = source or base population
C/C̄ = prevalent cases/noncases
D/D̄ = incident cases/noncases or deaths/survivors
E/Ē = subjects with/without the exposure
s = random sampling

Figure 1. Diagrammatic Representation of a Cohort Study

Retrospective cohort studies, also known as historical cohort studies, require the identification of individuals based on their past exposure and the reconstruction of their disease experience up to a defined point in time. Retrospective cohort designs are often useful for examining the effects of occupational exposures. Unlike prospective cohort studies, they often rely on already-existing records and may consequently be completed in less time and with lesser cost than a prospective cohort study. Retrospective cohort studies do, however, share some of the same difficulties as prospective studies, including difficulties in the ascertainment and measurement of extraneous relevant characteristics (confounding variables), and difficulties tracing individuals through time. Despite the difficulties inherent in cohort designs, cohort studies offer a major benefit: the ability to calculate incidence rates for the exposed and the unexposed groups.

Case-control Studies

Unlike cohort studies which follow individuals through time after classifying them based on their exposure status, case-control studies require the classification of individuals on the basis of their current disease status and then examine their past exposure to the factor of interest (see figure 2). Case-control design has been used to investigate disease outbreaks (Dwyer, Strickler, Goodman, & Armenian, 1994); to identify occupational risk factors (Checkoway & Demers, 1994); in genetic epidemiologic studies (Khoury & Beaty, 1994); for indirect estimation in demography (Khlat, 1994); to evaluate vaccination effectiveness and vaccine efficacy (Comstock, 1994); to evaluate treatment and program efficacy (Selby,

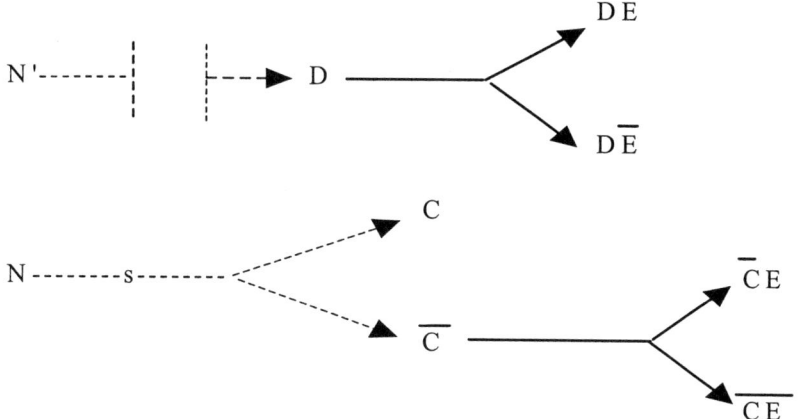

N = source or base population
$C/\bar{C}$ = prevalent cases/noncases
$D/\bar{D}$ = incident cases/noncases or deaths/survivors
$E/\bar{E}$ = subjects with/without the exposure
s = random sampling

Figure 2. Diagrammatic Representation of a Case-Control Study

1994); and to evaluate the efficacy of screening tests (Weiss, 1994). Case-control studies are most useful in evaluating risk factors for rare diseases and for diseases of rapid onset. With diseases of slow onset, it may be difficult to ascertain whether a particular factor contributed to disease causation or arose after the commencement of the disease process (Kelsey, Thompson, & Evans, 1986).

The conduct of a case-control study requires selection of the cases (the diseased group) and the controls (the nondiseased group) from separate populations. It should be obvious that prior to selecting the cases, the investigator must define what constitutes a case conceptually. This is not as simple a task as it first appears, particularly when the disease condition is a new and relatively unstudied entity (Lasky & Stolley, 1994). Causal inference is possible only if one assumes that the controls are "representative of the same candidate population . . . from which the cases…developed…." (Kleinbaum, Kupper, & Morgenstern, 1982: 68). Consequently, the selection of appropriate cases and controls is crucial to the validity of the study.

Methods and criteria for the selections of appropriate cases and controls have been discussed extensively in the literature and will only be summarized here. Cases are often selected from patients seeking medical care for the condition that is being investigated. It is preferable to include as cases individuals who have been recently diagnosed with the illness rather than individuals who have had the disease for an extended period of time, in order to discriminate between exposure that occurred before disease onset and exposure that occurred after. Other sources of cases include disease registries, drug surveillance programs, schools, and places of employment (Kelsey, Thompson, & Evans, 1986).

Controls must be "representative of the same base experience" as the cases (Miettinen, 1985). "[T]he control series is intended to provide an estimate of the exposure rate that would be expected to occur in the cases if there were no association between the study disease and exposure"(Schlesselman, 1982: 76). It is important to recognize that controls are theoretically continuously eligible to become cases. An individual who is initially selected as a control and who later develops the disease(s) under study, thereby becoming a case, should be counted as both a case and a control (Lubin & Gail, 1984). Controls are frequently selected from probability samples of the population from which the cases arose; from patients receiving medical care at the same facilities as the cases, but for conditions unrelated to the cases' diagnoses; or from neighbors, friends, siblings or coworkers of the cases (Kelsey, Thompson, & Evans, 1986). Dead controls may also be used in studies where the researcher wishes to compare individuals who died from one cause with individuals who died from other causes (Lasky & Stolley, 1994).

Case-control studies are valuable because they permit the evaluation of a range of exposures that may be related to the disease under investigation. They are generally less expensive to conduct than cohort studies, in part because fewer people are needed for the study. However, it may be difficult to determine individuals' exposure status (Rothman, 1986).

Cross-Sectional Studies

Unlike either cohort studies or case-control studies, exposure and disease status are measured at the same point in time in cross-sectional studies (see figure 3). This approach results in a serious limitation, in that it may be difficult to determine whether the exposure or the disease came first, since they are both measured at the same time. Additionally, because cross-sectional studies include prevalent cases of a disease, *i.e.,* new cases and already-existing cases, a higher proportion of cases will have had the disease for a longer period of time. This may be problematic if people who die quickly or recover quickly from the disease differ on important characteristics from those who have the disease over a long period of time. Too, individuals whose disease is in remission may be erroneously classified as nondiseased (Kelsey, Thompson, & Evans, 1986).

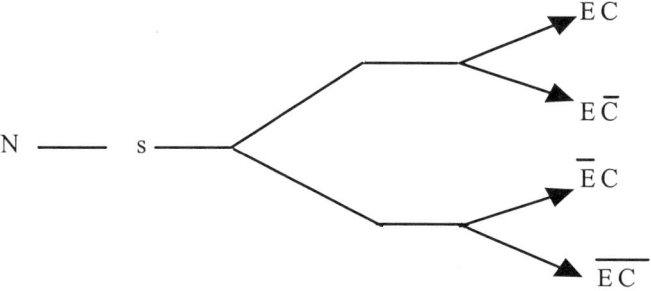

N = source or base population
$C/\overline{C}$ = prevalent cases/noncases
$D/\overline{D}$ = incident cases/noncases or deaths/survivors
$E/\overline{E}$ = subjects with/without the exposure
s = random sampling

Figure 3. Diagrammatic Representation of a Cross-sectional Study

Ecological Studies

In the study designs previously discussed, the individual was the unit of observation. Ecological designs, however, utilize a group of people, such as census tract data, as the unit of observation (Rothman, 1986). An example of an ecological study would be an examination of oral cancer rates against the use of chewing tobacco in each state. Ecologic studies are often conducted to observe geographic differences in the rates of a specific disease or to observe the relationship between changes in the average exposure level and changes in the rates of a specific disease in a particular population. They are useful in generating etiologic hypotheses and for evaluating the effectiveness of a population intervention. Because data is available only at the group level, however, inferences from the ecological analysis to individuals within the groups or to individuals across groups may be seriously flawed (Morgenstern, 1982). This often results from an inability to assess and measure extraneous factors on an individual level that may be related to the disease and the exposure under examination. See Figure 4.

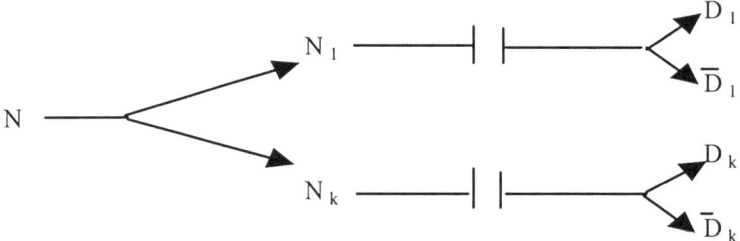

N = source or base population
C/C̄ = prevalent cases/noncases
D/D̄ = incident cases/noncases or deaths/survivors
E/Ē = subjects with/without the exposure
s = random sampling

Figure 4. Diagrammatic Representation of an Ecological Study

INTERPRETING THE RESULTS

Statistical Significance and Confidence Intervals

In testing an hypothesis of association between exposure and disease, the investigator often begins with the hypothesis that there is no association between the exposure and the disease (null hypothesis). If the data do not support the null hypothesis, it can be rejected. The investigator must specify the level of statistical significance (alpha) which will indicate that any association that is found is unlikely to have occurred by chance. Although this alpha level is usually set at .05, this is completely arbitrary (Rothman, 1986). It is important to remember that statistical significance does not equate to clinical significance; a result may be clinically significant even in the absence of statistical significance.

A "p-value" is a statistic used to test the null hypothesis. It refers to the probability that the data will depart from an absence of association, to an extent equal to or greater than that observed, by chance alone, assuming that the null hypothesis is true. A lower p-value indicates a higher degree of inconsistency between the null hypothesis of no association and the data. Stated more simply, the p-value is the probability that we will make a mistake and reject the hypothesis of no association when, in fact, it is true. We want the p-value to be very small. The smaller the p-value, the more certain we are that the null hypothesis is not true.

An alpha, or Type I, error occurs when the null hypothesis is erroneously rejected, *i.e.*, it is true and it is rejected as false. If the p-value was set to .05, an alpha error will occur approximately five percent of the time. Conversely, a beta, or Type II, error will occur if the null hypothesis is false and is not rejected.

Reliance on p-values has been criticized because p-values fail to provide information about the magnitude of an effect estimate or its variability (Rothman,

1986). Rothman has urged, instead, the use of interval estimation (confidence intervals). A point estimate, which is a single best estimate of the parameter, *e.g.,* an effect measure such as an odds ratio, is derived from the data. A confidence interval, equivalent to one minus the alpha level, is calculated around the point estimate. As an example, if the alpha level is .05, the confidence interval to be constructed is a 95% confidence interval.

Internal And External Validity

The scientific validity of a study depends on its internal validity and its external validity, or generalizability. Internal validity refers to the validity of the inferences drawn that relate to the study participants. External validity refers to the validity of the inferences drawn as they relate to groups other than the study population, such as all adults, or all children with a particular disease.

Internal Validity

Biases in research can impact on a study's internal validity by affecting the accuracy of measurement. The *Dictionary of Epidemiology* defines "bias" as "any trend in the collection, analysis, interpretation, publication or review of data that can lead to conclusions that are systematically different from the truth."(Last, 1988: 13-14). There are many types of bias, but they are often classified into three general categories: selection bias, information bias, and confounding.

Selection bias may result from flaws in the procedures utilized to select participants for the study. These flaws lead to a distortion in the estimate of effect (Kleinbaum, Kupper, & Morgenstern, 1982).

Information bias results in a distortion of the effect estimate due to misclassification of the research participants on one or more variables. The misclassification may result from measurement errors or recall bias. Misclassification is said to be nondifferential if the misclassification on one axis (exposure or disease) is independent of misclassification on the other axis. Differential misclassification occurs when misclassification on one axis is not independent of misclassification on the other axis (Rothman, 1986). Differential misclassification can result in an over- or underestimation of the effect measure (Copeland, Checkoway, McMichael, & Holbrook, 1977). Recall bias is one form of differential misclassification that may occur in a case-control study that relies on participants' memories of their exposure experiences. Memory may differ between the exposed cases and the nondiseased controls for a variety of reasons. As an example, particular exposures may become more significant to the cases in retrospect because of an attempt to identify a cause or reason for the illness.

Confounding can occur if the exposure of interest is closely linked to another variable and to the disease of interest. (See figures 5-19.) For instance, if one were to study the association between alcohol use and a specific form of cancer without collecting data on levels of smoking, the results would be confounded if that form of cancer were associated with tobacco usage.

Key: C=covariate D=disease E=exposure

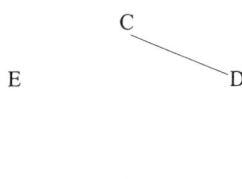

Figure 5. Factors C and D are associated in the study population, but not necessarily in the base population. This may result from selection procedures In this situation, C is not a confounder.

Figure 6. The C-D association results from the effect of D on C. C is not a confounder.

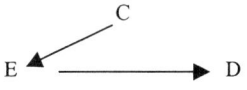

Figure 7. The effect of C on D in the base population is not independent of the exposure. C is not a confounder.

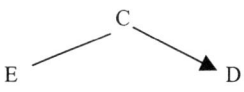

Figure 8. C is associated with exposure status in the base population and C is risk factor for D in the unexposed base population. C is a confounder.

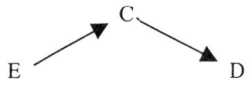

Figure 9. C is an intermediate (intervening) variable in the causal pathway between E and D. C is not a confounder.

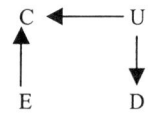

Figure 10. Both C and D are affected by the same unmeasured risk factor U and C is affected by E. C is not a confounder.

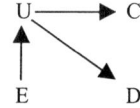

Figure 11. Both C and D are affected by another unmeasured risk factor, U. U is affected by E. C is not a confounder.

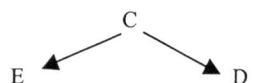

Figure 12. C is a risk factor for both E and D. C is a causal confounder.

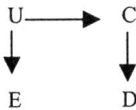
Figure 13. U is a causal confounder. C is in the causal pathway between U and D. By controlling for C, all confounding due to U will be eliminated, if there are no measurement errors.

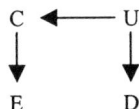
Figure 14. U is a causal confounder. C is in the causal pathway between U and E. By controlling for C, all confounding due to U will be eliminated, if there are no measurement errors.

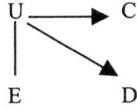
Figure 15. U is a causal or proxy confounder. C is associated with U. C is not in every causal pathway between U and E or between U and D. Controlling for C will will not eliminate confounding by U.

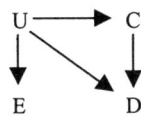
Figure 16. U is a causal or proxy confounder. C is associated with U, but is not in every causal pathway between U and E or U and D. Controlling for C could result in either increased or decreased bias because the direct and indirect effects of U on D could be in opposite directions.

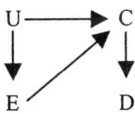
Figure 17. C is a proxy for the unmeasured confounder U. Confounding due to U will be eliminated by controlling for U if C is an intermediate variable and U is a confounder. If we have no information on U, C is a proxy confounder in addition to being an intermediate variable. Our estimate of E will be biased whether or not we control for C.

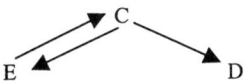
Figure 18. C is a time-dependent variable and is both a confounder and an intermediate between E and D. Conventional statistical methods will not produce an unbiased estimate of the effect of E.

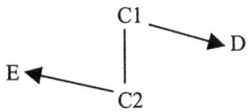
Figure 19. C2, a confounder, is a proxy for C1, a confounder. C1 and C2 are redundant confounders. Analysis must control for either C1 or C2 to eliminate bias.

To be a confounder, a factor (1) "must be associated with both the exposure under study and the disease under study ..." (2) must "be associated with the exposure among the source population for cases;" and (3) may not be "a step in the causal chain between exposure and disease...." (Rothman, 1982). It is important to note that, just as with exposure and disease status, confounders are subject to misclassification. Under some circumstances, such misclassification can seriously distort the results (Greenland & Robins, 1985).

Confounding is often confused with effect modification. Unlike confounding, which refers to a bias in the estimate of effect that results from a lack of comparability between the groups in the study, effect modification refers to a heterogeneity of effect. As an example, if the effect of smoking on cervical cancer differs by ethnicity, we would say that ethnicity in this case is an effect modifier. Ethnicity could also, however, be a confounder if there is an association between ethnicity and smoking among the controls.

External Validity

Scientific generalization is "the process of moving from time- and place-specific observations to an abstract universal statement" (Rothman, 1982: 96). As an example, the applicability to women of the results of clinical trials relating to heart disease and HIV in men, for example, has been called into question (Mastroianni et al., 1994).

Strategies to Increase Validity

A variety of options are available to control for potential bias. One such option, randomization, was discussed previously in the context of clinical trials and that discussion will not be repeated here. Other options include restriction, matching, stratification, and mathematical modeling.

Restriction is used in the design phase of a study to limit inclusion in the study to individuals who meet certain predetermined eligibility criteria. In this way, individuals who possess certain characteristics may be excluded, thereby reducing the potential of bias due to the presence of those extraneous factors (Gray-Donald & Kramer, 1988). As an example, investigators studying the etiology of HIV-associated dementia would exclude from the study individuals who possess other risk factors for dementia or dementia-like conditions, such as current alcohol or drug abuse or certain forms of mental illness. As a technique, restriction is relatively inexpensive and facilitates the analysis and interpretation of the results. However, generalization to populations other than those included in the study may not be valid. Additionally, restriction may not control completely for confounding (Kleinbaum, Kupper, & Morgenstern, 1982). Restriction may also raise ethical issues due to the routine exclusion of specific groups, such as women, from participation in research studies (Mastroianni et al., 1994).

"Matching" refers to the selection of a comparison group that is comparable to the study group with respect to certain prescribed characteristics that could potentially confound or bias the results, such as age or sex. Matching reduces or eliminates the potential variability between the study and comparison groups with respect to the "matched" variables. Consequently, the factors selected as the basis for matching should not be factors of interest that the investigator wishes to examine further (Kelsey, Thompson, & Evans, 1986). For example, an investigator may be interested in factors that affect the quality of care received by HIV-infected patients. If the investigator matches on insurance status, he or she will be unable to investigate the effect that insurance has on whether or not the patient receives a particular treatment or undergoes a particular procedure. Although matching can be used with cohort, case-control, and cross-sectional designs (Kelsey, Thompson, & Evans, 1986), it is most often used in case-control designs. Matching is often not feasible in cohort and cross-sectional studies due to lack of information on the potentially confounding factors (Kelsey, Thompson, & Evans, 1986).

The variables on which cases and controls are to be matched must be thought to be related to both disease and exposure, *i.e.*, they must be confounders that must be controlled for in the design or the analysis. Additionally, matching should not be unduly costly (Kelsey, Thompson, & Evans, 1986). Cost may become a particular issue if matched controls are difficult to locate due to the closeness of the matching that is required (Smith, 1983). Matching brings several advantages, including the ability to more adequately control for confounding variables, the ability to obtain time comparability between cases and controls for exposures that may vary over time, and a gain in statistical power (Wacholder, Silverman, McLaughlin, & Mandel, 1992).

Stratification means dividing the analysis into two or more groups, such as males and females, to permit separate analysis of each. Stratified analysis can be used in cohort, case-control, and cross-sectional studies. This technique requires that levels of the confounding variables be defined and the exposure-disease association estimated within each level. Stratification is appropriate if (1) there is a sufficient number of individuals in each level or strata; (2) the choice of control variables is appropriate; and (3) the definition of levels for each of the confounding variables is appropriate (Kleinbaum, Kupper, & Morgenstern, 1982). As an example, smoking could be a confounding variable in a study examining the relationship between asbestos exposure and certain lung diseases. Consequently, the investigator would want to stratify the analysis by level of smoking, *e.g.*, lifetime levels of smoking.

Mathematical modeling involves the use of analyses that relate exposure, outcome, and extraneous variables. These analyses are said to be multivariate because they include multiple factors in the model. In cohort studies, the illness or disease outcome or status is often the dependent variable, while either the exposure status or disease outcome may be the dependent variable in case-control studies. Mathematical modeling has many advantages, including ease of use with small numbers; the ability to predict individual risk; and the ability to use this technique with continuous variables and with multiple exposure variables.

Mathematical modeling may also entail several disadvantages. All models require that certain assumptions be made about the data prior to the application of the model. The selection of the model requires an evaluation of these underlying assumptions. If the assumptions do not hold, another modeling technique must be used. Mathematical modeling may also present difficulty in the interpretation of the results (Kleinbaum, Kupper, & Morgenstern, 1982).

SAMPLING AND SAMPLE SIZE

Many epidemiologic studies utilize data from already existing data sources, often collected on a routine basis, such as data from various disease registries or hospital discharge records. This type of data is known as secondary data. It is crucial that these data sources be as accurate as possible, *e.g.,* that minimal misclassification exists.

Most epidemiologic studies, however, rely on primary data collected from the original source, such as individuals with a particular disease or exposure of interest. This often requires the identification of the individuals who will participate, the development of questionnaires, and/or the conduct of interviews. In most cases, a sample of such individuals must be used, rather than an entire population. Reliance on a sample of the population helps to reduce study costs and, to some extent, may increase the accuracy of the measurements since more time can then be focused on fewer people.

When assembling a sample population, the sampling unit will depend on the particular study. Most often, the unit will be an individual or a household, although it may also be a neighborhood center, school, or other entity. A listing of the sampling units constitutes the sampling frame.

Sampling Techniques

The manner by which sampling is conducted is critical. Appropriate sampling techniques will reduce random error, which can reduce the precision of the epidemiologic measurements that are obtained (Rothman, 1986). Probability sampling is one method of reducing random error. Probability sampling refers to sampling by which "each sampling unit has a known, nonzero probability of being included in the sample" (Kelsey, Thompson, & Evans, 1986). There are various techniques available for probability sampling, including simple random sampling, systematic sampling, stratified sampling, cluster sampling, and multistage sampling. Other sampling techniques, such as snowball sampling, are available for use in situations where probability sampling is not possible.

Simple Random Sampling

With simple random sampling, each sampling unit in the eligible population has an equal chance of being included in the study. In order to conduct random sampling, the investigator must know the complete sampling frame, *i.e.,* everyone who is potentially eligible to be in the study. Sampling can occur with or without replacement. With replacement sampling, selected sampling units, such as the individuals selected for participation in the study, are returned to the pool from which the sample is being taken. Most epidemiologic studies rely on sampling without replacement, which yields more precise estimates.

Systematic Sampling

With systematic sampling, sampling units are selected from regularly spaced positions from the sampling frame, such as every tenth patient admitted to an inpatient facility. Systematic sampling is relatively easy to implement and does not require *a priori* knowledge of the sampling frame.

Stratified Sampling

Stratified sampling requires that the population be divided into predetermined strata. Within each strata, the sampling units share particular characteristics, such as sex. The study participants are then selected by taking a random sample from within each strata. Stratified sampling is particularly useful to ensure that all subgroups of interest are represented in the study population. Stratified sampling may also yield more precise estimates of the population parameters since the overall variance is based on the within-stratum variances (Kelsey, Thompson, & Evans, 1986).

Disproportionate stratification refers to the disproportionate sampling of strata, such as specific neighborhoods, which contain high concentrations of the population of interest. For instance, if an investigator were interested in studying the effect of culture on nutritional intake, he or she might oversample certain groups to ensure that there is a sufficient number of such individuals in the study to be able to analyze the data. This technique results in unequal selection probabilities for members of the different strata, thereby requiring weighting adjustments in the analysis of the data (Kalton, 1993).

Cluster Sampling

Cluster sampling involves the selection of clusters from the population. Observations are then made on each individual within a cluster. As an example, one may wish to identify certain neighborhoods (clusters) and then sample all households within those selected neighborhoods.

Multistage Sampling

Multistage sampling is similar to cluster sampling in that it first requires the identification of primary sampling units, such as neighborhoods. Unlike cluster sampling, however, multistage sampling utilizes a sample of secondary units with each primary unit, rather than a sample of all of the secondary units. As an example, multistage sampling would require the sampling of households within the selected neighborhoods, such as every tenth house in the selected neighborhoods, rather than reliance on all of the households within each selected neighborhood (Kalton, 1993).

Location Sampling

Location sampling refers to the selection of participants through recruitment at locations and times when numbers of the target population are expected to be high. Locations could include, for instance, bars, grocery stores, bookstores, or churches, depending upon the population to be sampled. This type of sample is generally considered a convenience sample.

Time/space sampling is conducted at specified locations and times when population flows are expected to occur, such as voting booths on election day. A sampling frame consisting of time/location combinations is constructed and a sample of individuals is then selected from these selected sampling units. This method of sampling produces a probability sample of visits, rather than individuals (Kalton, 1993).

Snowball Sampling

Snowball sampling is based on the premise that members of a particular, *i.e.*, rare, population know each other. Individuals are identified within the targeted rare population. These individuals are asked to identify other individuals within the same target group. This technique can be used to generate the sample ("snowball sampling") or, alternatively, to construct a sampling frame for the rare population, from which the sample is then selected. For example, if an investigator wished to examine the prevalence of needle sharing behaviors among injection drug users, snowball sampling would permit the investigator to identify eligible participants by relying on previously identified eligible participants. This could be more efficient than attempting to identify individuals through hospitals or clinics. Snowball sampling is a nonprobability sampling procedure. Snowballing for frame construction does not suffer from this weakness, but carries the possibility that socially isolated members of the rare population will be missing from the frame (Kalton, 1993).

Sample Size and Power

The epidemiologist is concerned not only with the appropriateness of the sampling method used, but also with the size of the sample. Increasing the size of the sample may reduce random error and increase precision.

Power calculations are often utilized to determine the requisite sample size. "Power" has been defined as "the probability of detecting (as 'statistically significant') a postulated level of effect" (Rothman, 1986). In order to calculate power, one must specify the (statistical) significance, or alpha level; the magnitude of the effect that one wishes to detect, such as an odds ratio of 2; the sample size of the exposed group in a cohort or cross-sectional study or the sample size of the diseased group in a case-control study; and the ratio of the size of the comparison group to the size of the exposed group in a cohort or cross-sectional study or to the size of the diseased group in a case-control study (Kelsey, Thompson, & Evans, 1986).

An alternative approach to assessing the adequacy of the sample size is to calculate the requisite size using one of the accepted sample size formulas. These formulas require that the investigator specify the level of statistical significance (alpha error); the chance of missing a real effect (beta error); the magnitude of the effect to be detected; the prevalence of the exposure in the nondiseased or the disease rate among the unexposed; and either the ratio of the exposed to the unexposed or the ratio of the cases to the controls (Rothman, 1986). Reliance on such formulas has been criticized because they create the illusion of a boundary between an adequate and inadequate sample size when, in fact, the variables specified to determine the sample size are often set either arbitrarily or by relying on estimates (Rothman, 1986).

Meta-analysis offers "[a] quantitative method of combining the results of independent studies (usually drawn from the published literature) and synthesizing summaries and conclusions which may be used to evaluate therapeutic effectiveness, plan new studies, etc.... "(Olkin, 1995: 133). Meta-analysis refers to not only the statistical combination of these studies, but also to "the whole process of selection, critical appraisal, analysis, and interpretation...."(Liberati, 1995: 81). Because meta-analysis permits the aggregation of studies' results, it may be useful in detecting effects that have been somewhat difficult to observe due to the small sample size of individual studies. Its use, however, is not uncontroversial, and various approaches to meta-analysis have been subject to criticism, including the use of quality scores for the aggregation of studies of both good and poor quality and over-reliance on p-values (Olkin, 1995). Additionally, because "[t]he validity of a meta-analysis depends on complete sampling of all the studies performed on a particular topic"(Felson, 1992: 886), the results of meta-analysis may be biased due to various forms of sampling bias, selection bias, or misclassification (Felson, 1992). The establishment of more or less rigid inclusion and exclusion criteria can impact heavily on the results of a meta-analysis.

References

Carter, K.C. (1985). Koch's postulates in relation to the work of Jacob Henle and Edwin Klebs. *Medical History, 29*, 353-374, 356-357.

Checkoway, H. & Demers, D.A. (1994). Occupational case--control studies. *Epidemiologic Reviews, 16*, 151-162.

Comstock, G.W. (1994). Evaluating vaccination effectiveness and vaccine efficacy by means of case control studies. *Epidemiologic Reviews, 16*, 77-89.

Copeland, K.T., Checkoway, H., McMichael, A.J. & Holbrook, R.H. (1977). Bias due to misclassification in the estimation of relative risk. *American Journal of Epidemiology, 105*, 488-495.

Coughlin, S.S. (1990). Recall bias in epidemiologic studies. *Journal of Clinical Epidemiology, 43*, 87-91.

Dwyer, D.M., Strickler, H., Goodman, R.A., & Armenian, H.K. (1994). Use of case-control studies in outbreak investigations, *Epidemiologic Reviews, 16*, 109-123.

Farrar, W.B. (1991). Clinical trials: Access and reimbursement. *Cancer, 67*, 1779-1782.

Feldman, F., Finch, M. & Dowd, B. (1989). The role of health practices in HMO selection bias: A confirmatory study. *Inquiry, 26*, 381-387.

Felson, D.T. (1992). Bias in meta-analytic research. *Journal of Clinical Epidemiology, 45,* 885-892.

Gore, S.M. (1981).Assessing clinical trials--why randomise? *British Medical Journal, 282,* 1958-1960.

Gray-Donald, K. & Kramer, M.S. (1988). Causality inference in observational vs. experimental studies: An empirical comparison, *American Journal of Epidemiology, 127,* 885-892.

Greenland, S. (1977). Response and follow-up bias in cohort studies. *American Journal of Epidemiology, 106,* 184-187.

Greenland, S. & Robins, J.M. (1985). Confounding and misclassification. *American Journal of Epidemiology, 122,* 495-506.

Hammond, E.C., Selikoff, I.J., & Seidman, H. (1979). Asbestos exposure, cigarette smoking, and death rates. *Annals of the New York Academy of Science, 330,* 473-490.

Hill, A.B. (1965). The environment and disease: Association or causation? *Proceedings of the Royal Society of Medicine, 58,* 295-300.

Hills, M. & Armitage, P. (1979). The two-period cross-over clinical trial. *British Journal of Clinical Pharmacology, 8,* 7-20.

Jooste, P.L., Yach, D., Steenkamp, H.J., Botha, J.L., & Rossouw, J.E. (1990). Drop-out and newcomer bias in a community cardiovascular follow-up study. *International Journal of Epidemiology, 19,* 284-289.

Kelsey, J.L., Thompson, W.D., & Evans, A.S. (1986). *Methods in Observational Epidemiology.* New York: Oxford University Press.

Khlat, M. (1994). Use of case-control methods for indirect estimation in demography. *Epidemiologic Reviews, 16,* 124-133.

Khoury, M.J. & and Terri H. Beaty, T.H. (1994). Applications of the case-control method in genetic epidemiology. *Epidemiologic Reviews, 16,* 134-150.

Kleinbaum, D.G., Kupper, L.L., & Morgenstern, M. (1982). *Epidemiologic Research: Principles and Quantitative Methods.* New York: Van Nostrand Reinhold.

Lasky, T. & Stolley, P.D. (1994). Selection of cases and controls," *Epidemiologic Reviews, 16,* 6-17.

Last, J.M., ed. (1988). *A Dictionary of Epidemiology,* 2nd ed. New York: Oxford University Press.

Liberati, A. (1995). 'Meta-analysis: statistical alchemy for the 21st century': Discussion. A plea for a more balanced view of meta-analysis and systematic overviews of the effect of health care interventions. *Journal of Clinical Epidemiology, 48,* 81-86.

Lubin, J.H. & Gail, M.H. (1984). Biased selection of controls for case-control analyses of cohort studies, *Biometrics, 40,* 63-75.

Miettinen, O.S. (1985). The case-control study: Valid selection of subjects. *Journal of Chronic Disease, 38,* 543-548.

Miles, & J. Evans (Eds.), *Demystifying Social Statistics* (pp. 87-109). London: Pluto Press.

Morabia, A. (1991). On the origin of Hill's causal criteria. *Epidemiology, 2,* 367-369.

Morgenstern, H. (1996, Spring). *Course Materials, Part II: Class Notes for Epidemiologic Methods II, Epidemiology 201B,* 54-82.

Morgenstern, H. (1982). Uses of ecologic analysis in epidemiologic research. *American Journal of Public Health, 72,* 1336-1344.

O'Brien, P.C. & Shampo, M.A. (1988). Statistical considerations for performing multiple tests in a single experiment. 5. Comparing two therapies with respect to several endpoints. *Mayo Clinic Proceedings, 63,* 1140-1143.

Olkin, I. (1995). Statistical and theoretical considerations in meta-analysis, quoting the National Library of Medicine, *Journal of Clinical Epidemiology, 48,* 133-146.

Popper, K.R. (1968). *The Logic of Scientific Discovery,* 3rd ed. rev. New York: Harper and Row.

Rosner, F. (1987). The ethics of randomized clinical trials. *American Journal of Medicine, 82,* 283-290.

Rothman, K.J. (1976). Causes, *American Journal of Epidemiology,104,* 587-592.

Rothman, K.J. (1986). *Modern Epidemiology.* Boston: Little, Brown and Company.

Sackett, D.L. (1979). Bias in analytic research. *Journal of Chronic Diseases, 32,* 51-63.

Sartwell, P.E. (1960). On the methodology of investigations of etiologic factors in chronic disease further comments. *Journal of Chronic Disease, 11,* 61-63.

Schlesselman, J.J. (1982). *Case Control Studies.* New York: Oxford University Press.

Selby, J.V. (1994). Case-control evaluations of treatment and program efficacy. *Epidemiologic Reviews, 16,* 90-101.

Senn, S.J. (1991). Falsification and clinical trials. *Statistics in Medicine, 10,* 1679-1692.

Shapiro, A.K. & Shapiro, E. (1997). *The Powerful Placebo: From Ancient Priest to Modern Physician.* Baltimore, Maryland: Johns Hopkins University.

Smith, P.G. (1983). Issues in the design of case-control studies: Matching and interaction effects. *Tijdschrift voor Sociale Gezondheidszorg, 61,* 755-760.

Sterling, T.D., Weinkam, J.J. & Weinkam, J.L. (1990). The sick person effect. *Journal of Clinical Epidemiology, 43,* 141-151.

Susser, M. (1986). The logic of Sir Karl Popper and the practice of epidemiology. *American Journal of Epidemiology, 124,* 711-718.

Susser, M. (1991). What is a cause and how do we know one? A grammar for pragmatic epidemiology. *American Journal of Epidemiology, 133,* 635-648.

Wacholder, S., Silverman, D.T., McLaughlin, J.K., & Mandel, J.S. (1992). Selection of controls in case control studies. III. Design options. *American Journal of Epidemiology, 135,* 1042-1050.

Weed, D.L. (1986). On the logic of causal inference. *American Journal of Epidemiology, 123,* 965-979.

Weiss, N.S. (1994). Application of the case-control method in the evaluation of screening, *Epidemiologic Reviews, 16,* 102-108.

21 Code of Federal Regulations section 312.21(a)(1998).

21 Code of Federal Regulations section 312.21(b)(1998).

21 Code of Federal Regulations section 312.21(c)(1998).

APPENDIX 2

BASIC LEGAL CONCEPTS

This section provides a review of basic concepts relating to the structure of our legal system, the derivation of law, and the procedures that may be relevant to legal actions.

Law is frequently classified into two domains, the public and the private. Public law encompasses law that is concerned with government or its relations with individuals and businesses. Public law is concerned with the definition, regulation, and enforcement of rights where an entity of the government is a party to the action. Public law derives from constitutions, statutes, and regulations and rules that have been promulgated by an administrative entity, such as a federal agency. For instance, the regulations of the Food and Drug Administration with respect to informed consent would be classified as public law.

Private law refers to law that regulates the relations between and among individuals and individuals and businesses. This includes actions relating to contracts, to property matters, and to torts. The primary sources of private law include statutes and judicial decisions.

Law is also classified into criminal and civil law. Criminal law deals with crimes. Even though a crime may have been committed against a person, for instance, when a person is robbed, the crime is said to have been against the state and it is the state (or federal government, depending upon the nature of the crime and the basis of the charge) that has the right to prosecute the accused individual or entity. Civil law is that law that refers to non-criminal public and private law.

SOURCES OF LAW

The sources of law can be thought of as being in an inverted pyramidal shape. At the very base of this inverted pyramid is the constitution. Everything above the constitution must be consistent with the principles enunciated in the constitution. Above the constitution are the statutes. As you move up the inverted pyramid, you find the regulations and the precepts that have been derived from cases heard by the court. At each level, the decisions and principles must be consistent with those of the previous levels. Although it would seem that the system is relatively unstable because the constitution, which forms the basis for everything else, is at the point of the pyramid, it is actually quite stable because everything else must remain in balance with the constitution.

The Constitution

The federal constitution has been called the "supreme law of the land." The Constitution is actually represents a grant of power from the states to the federal government because all powers not specifically delegated to the federal government are, pursuant to the terms of the Constitution, reserved to the states.

The Constitution allocates power among three branches of government. The legislative branch is charged with the responsibility and delegated the authority to make laws (statutes). The executive branch of the government is responsible for the enforcement of the laws, while the judicial branch is responsible for the interpretation of those laws.

There are 26 amendments to the main body of the Constitution. The first 10 of these amendments are known as the Bill of Rights. These encompass many of the rights with which people may be most familiar, such as freedom of speech and freedom of religion. It is important to remember, though, that these rights as delineated are in the federal constitution and as such apply to the federal, not state, government. The Fourteenth Amendment, however, provides specifically that no state may deprive any person of life, liberty, or property without the due process of law. The Amendment also provides that no state may deny equal protection to any person within its jurisdiction. Most of the rights that are enumerated in the Bill of Rights have been found by the Supreme Court to constitute due process, so that ultimately, these rights also apply to the states as well as to the federal government.

Each of the 50 states also has its own constitution. The state constitutions cannot grant to persons fewer rights than are guaranteed to them by the federal constitution. However, they may grant more rights than are provided for by the federal constitution.

Statutes

Statutes at the federal level are promulgated by Congress, consisting of the House of Representatives and the Senate. At the state level, the state legislatures, also consisting of two houses, are responsible for the promulgation of statutes. For example, Congress passed the laws which give the Food and Drug Administration and the Department of Health and Human Services their authority to make regulations. Judges are responsible for the interpretation of the statutes were there is a lack of clarity or where there is conflict between various statutory provisions.

Administrative Law

Administrative law is that law that is made by the agencies which comprise a part of the executive branch of government. Administrative law encompasses regulations, rules, guidelines, and policy memoranda. Examples of administrative agencies relevant to the health research context include the Food and Drug Administration, the National Institutes of Health, and the Department of Health and Human Services. Stated simplistically, an agency's regulations are developed and promulgated through a notice and comment procedures, whereby the proposed regulation is published in the *Federal Register*, which is available to the public for review. Following a mandated time period during which the promulgating agency may receive comments on its proposed regulation, the comment period will cease. After reviewing the comments and incorporating those that the agency deems appropriate, the agency will issue its final regulation. A similar process is followed on the state level.

Court Decisions

As indicated, judicial decisions must be consistent with statutes and the Constitution. The courts adhere to the doctrine known as *stare decisis*, meaning that they must look to past cases with similar facts and legal issues to resolve the cases that appear before them. In general, they are bound by decisions of all higher courts within the same jurisdiction. This will become clearer following a discussion of the structure of the legal system. For instance, all federal and state courts are bound by the decisions of the Supreme Court of the United States. All federal district courts are bound by the decisions of the federal Court of Appeal for the circuit in which the federal district court sits, but they are not bound by the decisions of a Court of Appeal for a different circuit. For instance, California sits in the Ninth Circuit. The federal district court for the southern district of California is bound by the decisions of the Ninth Circuit Court of Appeal, but is not bound by the decisions of the Fifth Circuit, which covers the geographic area encompassing such states as Texas and Louisiana.

Judicial decisions also follow the doctrine of *res judicata*. This means that once a case had been decided and all of the channels for appeal have been utilized, the party bringing the case may not bring it again.

THE STRUCTURE OF THE LEGAL SYSTEM

The state and federal court systems can be thought of as pyramids. At the very base of the pyramid are the lowest courts. At the mid-level of the pyramid sit the courts of first appeal and, at the pinnacle of the pyramid, sits the supreme court of the state or of the federal court system. Different states, however, name these various levels differently. For instance, the supreme court in California is known as the Supreme Court, but in Massachusetts it is known as the Massachusetts Supreme Judicial Court, and in New York it is called the Court of Appeals.

In the state court system, the lowest level courts are often divided into those that have limited jurisdiction and those that have general jurisdiction. Those with limited jurisdiction often hear cases involving less serious offenses and civil lawsuits that do not involve large sums of money. Courts of general jurisdiction may hear cases involving monetary sums over a specified amount or more serious matters. Courts of general jurisdiction are often divided into special courts due to the volume of cases and the need for specialized expertise. Examples of such specialized courts include juvenile court, family court, and family court.

The mid-level courts, or appellate courts, have the power to hear appeals from the decisions of the lower courts. This is known as appellate jurisdiction, as contrasted with original jurisdiction, which is the power to hear a case at its inception. The appellate courts may have original jurisdiction with respect to a limited range of cases. The state supreme court may hear appeals from the appellate courts.

The lowest tier on this pyramid in the federal system consists of the federal district courts. These courts hear cases involving crimes that arise under federal statutes, such as making false statements on a federal application. They have jurisdiction over cases in which the citizen of one state is suing a citizen of another state (diversity of citizenship case) if the amount in dispute is greater than a specified minimum. (State courts may also hear cases in which a citizen of one state is suing a citizen of another state. This is known as concurrent jurisdiction. Not infrequently, the party who did not file the original lawsuit may ask to have the case removed to federal court.) The federal district courts may also hear cases arising under the federal constitution and cases arising under federal statutes.

Appeals from the decisions of the district courts are made to the federal Court of Appeal having jurisdiction over the circuit in which the district court sits. There are 13 Court of Appeal. Twelve of these are for named circuits, one is for cases arising in the District of Columbia, and the 13th is for the Federal Circuit, which has jurisdiction over claims that are exclusively within the domain of federal law, patent and trademark law.

The Supreme Court hears appeals from the Courts of Appeal. However, in most situations, there is no automatic right to appeal to the Supreme Court. Rather, the Supreme Court chooses the cases that it will hear. Request to have an appeal heard is made through a *writ of certiorari*, which is a petition to file an appeal.

Apart from the judicial system, some agencies may have the power to resolve cases administratively. For instance, the Office of Research Integrity has the authority to investigate and adjudicate allegations of scientific misconduct. Appeals proceed to the Department Appeals Board and, from there, to court if necessary.

CIVIL PROCEDURE

A lawsuit is commenced through the filing of a complaint by a party to the lawsuit. The complaint must, in general, state the nature of the claim, the facts to support the claim, and the amount in controversy. The defendant will be served with a copy of the complaint, together with a summons. The summons indicates that the defendant must respond to the complaint in some fashion within a specified period of time or the plaintiff will win the lawsuit by default.

The defendant will then answer the complaint, and will admit, deny, or plead ignorance to each allegation of the complaint. The defendant may also file a countersuit against the plaintiff or against a third party. The defendant may also ask that the court dismiss the plaintiff's action, claiming that the court has no jurisdiction to entertain the case or that the plaintiff failed to state a cause of action.

Following the initiation of the lawsuit and the answer by the defendant, there will be a period of discovery, during which each party to the action will have the opportunity to gather additional facts to support its case, to identify expert witnesses that the other side may call, and to identify weaknesses in the opposing party's case. Discovery may include depositions, written interrogatories, the production of documents, a request for a mental or physical examination, and a request for admissions. Those that are most relevant to the health research context are depositions, written interrogatories, a request for the production of documents, and a request for admissions.

The trial itself consists of numerous stages:
1. the opening statement of the plaintiff,
2. the opening statement of the defendant,
3. the presentation of direct evidence by the plaintiff, with cross-examination of each witness by the defendant, re-direct by the plaintiff, and re-cross by the defendant,
4. the presentation of direct evidence by the defendant, with cross-examination by the plaintiff, re-direct by the defendant, and re-cross by the plaintiff,
5. presentation of rebuttal evidence by the plaintiff,
6. presentation of rebuttal evidence by the defendant,
7. plaintiff's argument to the jury,
8. defendant's argument to the jury,
9. plaintiff's closing argument to the jury,
10. instructions from the judge to the jury, and
11. jury deliberation and verdict.

PROVING CAUSATION IN NEGLIGENCE ACTIONS

As indicated in chapter 5, in order to establish causation, the plaintiff must establish that the defendant owed a duty to the plaintiff, that that duty was breached, that the breach of that duty resulted in harm to the plaintiff (cause in fact), that there was a nexus between the defendant's conduct and the plaintiff's injury (proximate cause), and that the plaintiff is seeking damages. Legal jurisdictions differ with respect to whether a duty is owed to plaintiffs who may not be foreseeable. In some jurisdictions, if a duty is owed to anybody, it is owed to everybody. In other jurisdictions, duty is not owed to those who are unforeseeable. Even where, however, the plaintiff may be foreseeable, there may not be a duty owed depending upon the closeness of the connection between the defendant's conduct and the harm that the plaintiff suffered.

Establishing a Breach of a Duty

In order to establish that the defendant breached a duty of care, the plaintiff must provide proof of what actually happened and must demonstrate that the defendant acted unreasonably under the circumstances. In determining whether the conduct is unreasonable, the court may look at the balance between the risks and the benefits of the defendant's conduct. The risk refers to the severity of the harm that might occur as a result f the defendant's conduct and the probability that that harm will occur. In evaluating the benefit, the court may consider such things as the existence and availability of safer alternative methods, the costs of these alternative methods, and the social value attached to the defendant's conduct. The defendant's conduct will be considered unreasonable if a reasonable person in the defendant's position would have perceived in advance that the risks of the conduct outweighed its benefits.

Establishing Cause in Fact

There are several different rules, or standards, to determine whether the conduct of the defendant was responsible for the injuries suffered by the plaintiff. Different jurisdictions follow different rules. As a result, recovery by a plaintiff against a defendant may depend on the jurisdiction in which the harm occurred and the lawsuit is filed, even given the same facts.

The "but for" rule basically says that the plaintiff would not have been injured, but for the conduct of the defendant. This standard is essentially the legal equivalent of the deterministic model of causation in epidemiology, discussed in the previous appendix.

A second rule of causation is that of concurrent liability. For instance, if the plaintiff is injured through the actions of a defendant, together with the actions of a third party, and the plaintiff would not have been injured but for the concurrence of the actions of the defendant and the third party, then both the defendant and the third party will be said to be the actual cause of the injury.

The third rule of causation is that of the substantial factor. Assume that a plaintiff suffers an injury due to the conduct of the defendant and a third party. If the defendant's conduct was a substantial factor in bringing about the injury, he or she will be found to have caused the plaintiff's injury. This is the legal equivalent of Rothman's modified determinism, whereby multiple factors may combine to bring about a result. If this type of fact situation were to occur in a jurisdiction that utilized a "but for" rule, both the defendant and the third party would be found not to be liable because in neither instance could it be demonstrated that but for the conduct, the plaintiff would not have been harmed.

Some jurisdictions have accepted the principle of alternative liability. In such cases, there may be several defendants, each of whom committed the same conduct. However, it is impossible to determine whether it was the conduct of one or the other that resulted in harm to the plaintiff. This rule has led to the development of what is known as market share liability, which has been used in the context of a number of health-related cases.

For instance, consider the use of DES and the resulting harm. Various manufacturers may have made the product, but it may be impossible at the time that the lawsuit is filed for a plaintiff to know which manufacturer made the DES that she had ingested, and the DES manufactured by the various companies was essentially indistinguishable between those companies. New York has adopted the view that liability is related to the defendant's national market share of the product. The defendant will be found not to be liable only if it can show that it did not produce the product for the use that injured the plaintiff (*Hymowitz v. Eli Lilly*, 1989). In contrast, California finds that a defendant will be liable for the percentage of the plaintiff's injuries that is equal to the market share held by the company (*Sindell v. Abbott Laboratories*, 1980).

Some jurisdictions rely on a fifth rule, known as loss of a chance. In such instances, the plaintiff must demonstrate that he or she has lost something that he or she more likely than not would have retained or acquired, but for the conduct of the defendant. For instance, some courts have allowed a plaintiff to recover against the defendant where the plaintiff was physically hurt and now fears further harm, such as the development of cancer (*Mauro v. Raymark Industries, Inc.*, 1989).

Establishing Proximate Cause

The term "proximate cause" is actually a misnomer, because it refers not to cause, but to a policy decision: to which consequences of his or her conduct can the defendant's liability be extended? This determination is often made with reference to the foreseeability of the plaintiff, the foreseeability of the manner in which the breach of the duty occurred, and the foreseeability of the result. It is important to remember that, if there was no duty owed to the plaintiff, the issue of foreseeability is never reached. Other factors may also be considered, such as whether there were intervening acts that either extended the results of the defendant's conduct or combined with the defendant's conduct to produce the harm suffered by the plaintiff.

References

Hymowitz v. Eli Lilly & Co., 73 N.Y.2d 487, *cert. denied*, 493 U.S. 944 (1989).
Mauro v. Raymark Industries, Inc., 561 A.2d 257 (N.J. 1989).
Sindell v. Abbott Laboratories, 26 Cal. 3d 588, *cert. denied*, 449 U.S. 912, 1980.

INDEX

Acquired immune deficiency syndrome (AIDS) (See human immunodeficiency virus)
Administrative law, 240
Advance directives, 144-146
Advisory Committee on Human Radiation Experiments, 24
Advisory committees, 149
Advocacy, 162-163
African Americans
    and HIV conspiracy, 118-119
    and medical care, 3-8
    and prison experiments, 26
    and Reconstruction, 4-5
    and recruitment, 120
    and slavery, 2-4
Alsabti, 177-178
Altruism, 100
American Civil Liberties Union, 27
American Journal of Public Health, 154
American Journal of Respiratory and Critical Care Medicine, 113
American Journal of Respiratory Cell and Molecular Biology, 113
American Medical Association (AMA), 6, 24
Angelides, 183-185
Angell, 151
Anti-semitism, 11-14
Aristotle, 65
Arras, 46
Assent, 118, 135, 137-139, 153
Assurance of Compliance, 172, 173
Atlanta, 28-29
Atomic Energy Commission, 23
Auschwitz, 18
Australia, 97
Authorship
    and integrity in the review process, 157
    and morally tainted experiments, 149-151
    and participant confidentiality and privacy, 151-153
    and qualification as an author, 154
    and source of support, 155-157
Autonomy, 145, 158

Barrett v. Hoffman, 36
Battery, 190-191, 194
Baumrind, 72
Beauchamp, 63-65
Beecher, 122
Belmont Report, 36
Benefits, 86-88, 114, 116-117, 121, 161, 163,
Beneficence, 58, 59, 84
Bentham, 62
Bias
    conflict of interest, 100-105
    confounding, 229-232
    in study design, 219, 229-234
    information, 229
    selection, 229
Bioresearch Monitoring Program, 186
Blum, 53
Brady v. Hopper, 205
Breuning, 193
British Medical Journal, 154
Broad, 178
Buchenwald, 18

California, 203, 206-207, 243, 246
Cantwell, Jr., 37
Cancer, 4, 29, 30
Capacity, 79-80, 118, 130-142
Caplan, 156
Card, 54
Case
    definition, 46
    paradigm, 46
Case-control study design, 224-225
Casuistry, 45-47
Categorical Imperative, 60, 115
Causation
    epidemiology, 217-219
    legal, 191, 243-245
Cause in fact, 245-247

Centers for Disease Control, 171
Central Intelligence Agency, 29
Children, 21-22, 79-80, 124, 135-141
Childress, 63, 65
Cholera, 3
Clinical investigation, defined, 185
Clinical trial, 139-140, 159-160, 221-223
Clouser, 59
Cognitive impairment (see also mentally ill), 132-135
Cohort study design, 223-224
Cold War
    and radiation experiments, 20-25, 114
    historical origin, 19-20
Communication
    communicative-action model, 57-58
    information-transfer model, 57-58
    physician-patient, 57-58
    researcher-participant, 57-58
Communicative-action model, 57-58
Communitarianism, 47-48, 81, 114
Community, defined, 48
Community consultation, 149
Competence, 59
    and capacity, distinguished, 79
Confidentiality, 147-148
    certificate of, 209-211
    limitations of, 196-208
    mechanisms to enhance, 208-212
Conflicts of interest, 99-108, 161
    altruism, 100, 104
    disclosure, 108
    financial, 100-104
Confounding, 228-232
Connecticut, 198
Consent (see also informed consent), 1
Consequentialism, 63-64
Conspiracy theory, 119
Constitution, 242
Contract-based ethics, 64
Contracts, 102, 103, 105-106
Council on Ethical and Judicial Affairs, 101
Council on Scientific Affairs, 101
Cross-over design, 222-223
Cross-sectional study design, 226-227

Dachau, 18
Darwin, 9
Data
    disclosure, 159-160, 162-163
    fabrication, 175-185
    falsification, 175-185
    ownership, 105
    protection, 208-212
Data Safety Monitoring Board, 148
de Gobineau, 8
Debarment, 173-174
Deception
    and acceptability of, 72-73, 128-129
    and Tearoom Trade, 31-32, 72
Defense Atomic Support Agency, 22
Denmark, 177
Deontology, 60-61, 65, 77, 84, 114, 153, 178
Department of Health, Education and Welfare, 7
Departmental Appeals Board, 183-185
Depositions, 242
Determinism, 217-219
Dewey, 65-66
Diethylstilbesterol, 30, 158, 246
Discovery, 245-246
Donagan, 63
Doyal, 152, 153
Driskill, 202-203
Duke University Medical Center, 173
DuPont, 26
Duty to warn, 203-206

Earle v. Kuklo, 204
Englehardt, 156
Ethic of care, 51-53
Ethical pluralism, 115
Ethical relativism, 115
Ethical universalism, 115
Etzioni, 47

Eugenics, 8-11
Euthanasia, 14, 19
Exclusion criteria, 56, 76-77
Experiments, 15-32, 35, 219-220
    Candida albicans, 26
    contraceptive, 18
    dermatology, 26
    diethylstilbesterol, 30
    Fernald School, 21-22
    hepatitis, 30-31
    herpes simplex, 26
    herpes zoster, 26
    malaria, 18, 28-29
    Massachusetts Institute of Technology, 21-22
    Nazi, 18, 114, 151
    Patch Test, 26
    phototoxic drugs, 26
    radiation, 18
    ringworm, 26
    syphilis, 6-8, 14
    Tearoom Trade, 31-32
    University of Cincinnati, 22-23
    University of Chicago, 28, 30
    Vanderbilt Nutrition Study, 19-20
    Willowbrook, 30-31, 114

Feminine ethics, 48-51
Feminism
    liberal, 50
    Marxist, 50
    psychoanalytic, 50
    radical, 50
    socialist, 50-51
Feminist ethics, 48-58
Fernald School, 21-22
Finland, 177
Florida, 197-198, 202-203
Food and Drug Administration, 158, 185-188, 207
Freedman, 150
Freedom of Information Act, 171, 172, 201-203
Friter v. Iolab Corporation, 190

Gambia, 8, 119
Genocide, 8, 119

Gert, 59
Gilligan, 52-53
Great Migration, 5
Greene, 150-151
Guidelines for the Conduct of Health Research Involving Human Subjects, 98
Guidelines for Good Clinical Practice for Trials on Pharmaceutical Products, 71, 75, 90, 107

Harvard University, 30
Health Research Extension Act, 182
Helena Rubsenstein, 26
Helsinki Declaration33, 71, 74-75, 87, 88, 128, 147
Hill, 218-219
Hoagland, 54, 55
Holmesburg Prison, 25-27
Homosexuality, 31-32, 37
Hoseini v. United States, 205
Human immunodeficiency virus (HIV), 118-119, 148, 196, 206, 231
Humphreys, 31-32

In re Grand Jury Subpoena Dated January 4, 1984, 199-200
Inclusion criteria, 76-77
Information, 117, 127-130
Information bias, 229
Information-transfer model, 57-58
Informed consent, 7, 58, 113-149, 190
    elements, 58-59
    process, 142-146
    readability, 131-132
Institutional Ethics Committees, 97
Institutional Review Board, 91-96, 190
    deficiencies, 93-95
    function, 92-93, 95-96
    organization, 95
Institutional Review Committees, 98-99
International Committee of Medical Journal Editors, 151-152, 15

International Covenant on Civil and Political Rights, 33, 113
International Ethical Guidelines for Biomedical Research Involving Human Subjects, xii, 34, 71, 79, 82, 88, 90, 113, 115, 122, 128, 130, 137, 143-144
International Guidelines for Ethical Review of Epidemiological Studies, 34, 85, 90, 91, 100, 101, 106, 107, 121, 124, 126, 128, 154, 159, 161, 162
Invasion of privacy, 192

Jablonski v. United States, 205
Jews, 9-14, 18-19
Johnson & Johnson, 26
Jonsen, 45, 46, 47
Journal of Racial and Social Biology, 9
Justice, 36, 58, 64, 125-127, 179, 181

Kaimowitz v. Department of Mental Health for the State of Michigan, 93-94
Kaiser Wilhelm Institute for Anthropology, Human Genetics, and Eugenics, 10
Kant, 60
Kennedy, 25
Kligman, 26, 27
Koch, 217-218
Kohlberg, 52-53
Krugman, 30, 31
Kubie, 178
Kuzma, 179

Lamarck, 9-10
Law for the Protection of the Genetic Health of the German People, 13
Lehmann, 9
Lenz, 10, 11
Leonard v. Latrobe Area Hospital, 205
Lesbian ethics, 55
Leukemia, 29
Levine, 78, 122, 161
Lipari v. Sears, Roebuck & Co., 205

Loewy, 65
Luna, 150-151

Malaria, 4
Manslaughter, 194-195
Margolis, 103
Massachusetts Institute of Technology, 21
Massachusetts Task Force on Human Subject Research, 21
Matching, 231
Mathematical modeling, 232
McIntosh v. Milano, 204-205
Mendel, 9
Mengele, 10
Mentally ill, 79-81, 130-135
Mill, 51, 62
Mink v. University of Chicago, 190
Misrepresentation, 192-193
Misuse of human participants, 171-175
Modified determinism, 218
Moore v. Regents of the University of California, 191-192
Moral judgment, development of, 52-53
Motion to quash, 211
Munchausen's syndrome, 178
Murder, 194-195

National Bioethics Advisory Commission, 135
National Commission for the Protection of Human Subjects of Biomedical and Behavioral Research, 36, 127
National Council of Science and Technology, 98-99
National Health and Medical Research Council, 97
National Hospital Association, 5
National Institutes of Health, 23, 25, 36, 102, 140, 171, 193
National Medical Association (NMA), 5
National Research Act, 36
National Science Foundation, 175-176, 178

National socialism, 10-12
Nationalist Socialist Physicians' League, 11
Neglect, 197-199
Negligence, 191-193
Neisser, 14
Netherlands, 97-98
Nevada, 198
New England Journal of Medicine, 151
New York, 241
Nicholson, 53
Noddings, 53, 54
Nonmaleficence, 58, 59, 84, 103
Norway, 177
Notice of Opportunity for a Hearing, 187
Nuremberg Code, 24, 32-33, 74, 78, 82, 87, 113, 133

Office of Protection from Research Risks, 171-174
Office of Racial Policy, 13
Office of Research Integrity, 171, 180-183
Office of Scientific Integrity, 182
Office of Scientific Integrity Review, 182
Ohio, 29, 197, 199
Onek, 176
Origin of the Species, 9

Paradigm case, 46
Part 16 hearing, 187-188
Partner notification laws, 206-107
Patch test, 26
Peer review, 107
Pellagra, 4, 29-30
Pennsylvania, 197
Pharmaceutical companies
    and conflict of interest, 103
Physicians
    and Nazi cause, 11-14
    and slavery, 3-4
Placebo, 72, 84-86, 118, 222
Plagiarism, 175
Plato, 65
Ploetz, 9

Power, statistical, 235-236
Pozos, 150
Pragmatism, 65
Pregnancy, 212
Principlism, 45, 48, 58-50, 65, 84
Prisoners, 25-30, 123
Privacy, 208-209
Privacy Act, 208-209
Proximate cause, 245
Proxy consent, 113, 137
Public Health Service, 6, 107-108, 172-173, 179,180, 182
Public Health Services Act, 209
Purdy, 55

Quasi-experiment, 219

Racial hygiene, 9
Radiation, 18, 20-25, 26,
    and Cold War experiments, 20-25
    and Nazi experiments, 18
Ramsey, 136
Randomization, 219
Rawls, 64, 81
Recognition, 104
Reconstruction, 4-5
Records, defined, 201-202
Recruitment, 116-127
    barriers to, 118, 120, 125
    strategies, 120-127
Redmon, 136
Refugees, 79
Regulations on New Therapy and Human Experimentation, 15
Reich Citizenship Law, 13
Relational ethics, 53-54
Relativists, 33, 115
Relman, 150
Rennie, 150
Reporting laws, 196-199
Reproductive technology, 56
Research Ethics Committees, 97-98
Respect for persons; see autonomy, confidentiality
Rikkert, 150
Risks and benefits, 86, 87, 88
Roberts, 56

Rothman, 162
Rousseau, 51
Royalties, 103

Saenger, 22
Sample size, 234, 236-237
Sampling technique, 234-236
    cluster sampling, 235
    location sampling, 236
    multistage sampling, 235-236
    simple random sampling, 234-235
    snowball sampling, 236
    stratified sampling, 235
    systematic sampling, 235
Sandel, 47
Scientific misconduct, 175-185
    definition, 175-176
Scientific Office of Medical Health, 15
Selection bias, 219
Sen, 64
Sherwin, 49, 51, 55
Shimm, 106
Slavery, 2-9
Sloan-Kettering Institute, 29
Snider, 152
Soldiers, 79
Spece, 106
Stanton, 51
Stateville Penitentiary, 28
Sterilization Law, 13
Stigmatization, 119
Stock, 103
Stratification, 232
Study design, 71-74, 224-226
    case-control, 224-225
    clinical trials, 221-223
    cohort, 223-224
    crossover trials, 222-223
    cross-sectional, 226-227
    ecological, 227
    experimental, 219-220
    observational, 220-221
    quasi-experimental, 220-221
    restriction, 232
    stratification, 233
    validity, 229-234

Subpoena, 199-201, 211
Summerlin, 177
Suspension, 174-175
Sweden, 177
Syphilis, 1, 6-8, 14

Tarasoff v. Regents of the University of California, 203-206, 212
Tearoom Trade, 31-32, 72
Tetanus, 2
Texas, 198, 243
Thalassemia, 159-160
Thompson v. County of Alameda, 204
Tobacco, 102, 200
Tong, 49, 53-54
Tort, 189-192
    intentional, 189-191
    negligence, 189, 191-192
Total body irradiation, 22-23
Toulmin, 46
Toxic shock syndrome, 200
Tuberculosis, 2, 4, 5
Tulane University, 30
Tuskegee experiment, 208, 90, 114, 118
Typhoid, 1, 3
Typhus, 18
Typification, 45

Uganda, 79
Understanding, 59, 80, 117-118, 130-142
Universalists, 115
University of California, 191-192, 203-206
University of Cincinnati, 22-23
Utilitarianism, 61, 63, 65, 77, 81, 84, 153, 178
    and acts, 63
    and rules, 63

Validity, 227-230
Vanderbilt Nutrition Study, 19-20
Virtue ethics, 65
Voluntariness, 59, 78, 80, 115
Vulnerable participants, 77-83, 132-140
    children, 135-140

cognitively impaired, 132-135
definition and determination of status, 78
prisoners, 79
refugees, 79
soldiers

Wade, 178
Wagner, 10
Whistleblowing, 179
Whitbeck, 49
Willowbrook, 30-31, 114
Winkler, 59
Women
and exclusion from research, 56
and reproductive technology, 56
Writ of certiorari, 242

Yank, 150
Yellow fever, 1, 3